Monadologies

According to the received view, Kant's critical revolution put an end to the kind of metaphysics of which Leibniz's 'Monadology' is the example par excellence. This volume challenges Kant's claim by providing a far more nuanced version of philosophy's 'post-Kantian' tradition that spans from the late eighteenth to the early twentieth century and brings to light a rich tradition of new 'monadologists', many of whom have been unjustifiably forgotten by contemporary historians of philosophy. Through this complex dialogue, monadology is shown to be a remarkably fecund hypothesis, with many possible variations and developments. The volume's focus on monadology exposes the depth and breadth of the post-Kantian period in an original and previously unexplored way and opens up numerous avenues for future research. Crucially, however, this volume not only shows that monadological metaphysics *did* continue after Kant but also asks the critical question of whether it *should* have done so. Consequently, the question of whether monadological metaphysics could also have a *future* is shown to be relevant in a way that was previously almost inconceivable.

This book was originally published as a special issue of the *British Journal for the History of Philosophy*.

Jeremy Dunham is Assistant Professor in Philosophy at the University of Durham, UK.

Pauline Phemister is Professor of the History of Philosophy at the University of Edinburgh, UK.

Monadologies

Edited by
Jeremy Dunham and Pauline Phemister

LONDON AND NEW YORK

First published 2019
by Routledge
2 Park Square, Milton Park, Abingdon, Oxon, OX14 4RN, UK

and by Routledge
711 Third Avenue, New York, NY 10017, USA

Routledge is an imprint of the Taylor & Francis Group, an informa business

© 2019 British Society for the History of Philosophy

All rights reserved. No part of this book may be reprinted or reproduced or utilised in any form or by any electronic, mechanical, or other means, now known or hereafter invented, including photocopying and recording, or in any information storage or retrieval system, without permission in writing from the publishers.

Trademark notice: Product or corporate names may be trademarks or registered trademarks, and are used only for identification and explanation without intent to infringe.

British Library Cataloguing-in-Publication Data
A catalogue record for this book is available from the British Library

ISBN13: 978-0-8153-6220-3

Typeset in Times New Roman
by codeMantra

Publisher's Note
The publisher accepts responsibility for any inconsistencies that may have arisen during the conversion of this book from journal articles to book chapters, namely the possible inclusion of journal terminology.

Disclaimer
Every effort has been made to contact copyright holders for their permission to reprint material in this book. The publishers would be grateful to hear from any copyright holder who is not here acknowledged and will undertake to rectify any errors or omissions in future editions of this book.

Contents

Citation Information	vii
Notes on Contributors	ix
Introduction: Monadologies: An Historical Overview Jeremy Dunham and Pauline Phemister	1
1 Reconciling Leibnizian Monadology and Kantian Criticism Richard Mark Fincham	11
2 Herbart's Monadology Frederick Beiser	34
3 Bolzano's Monadology Peter Simons	52
4 From Habit to Monads: Félix Ravaisson's Theory of Substance Jeremy Dunham	63
5 Reviving Spiritualism with Monads: Francisque Bouillier's Impossible Mission (1839–64) Delphine Antoine-Mahut	84
6 Learning from Leibniz: Whitehead (and Russell) on Mind, Matter and Monads Pierfrancesco Basile	106
7 British Idealist Monadologies and the Reality of Time: Hilda Oakeley Against McTaggart, Leibniz, and Others Emily Thomas	128

CONTENTS

8 Heidegger on the Being of Monads: Lessons in Leibniz
   and in the Practice of Reading the History of Philosophy   147
   Paul Lodge

9 Five Figures of Folding: Deleuze on Leibniz's
   Monadological Metaphysics   170
   Mogens Lærke

10 Leibniz's Monadological Positive Aesthetics   192
   Pauline Phemister and Lloyd Strickland

   Index   213

Citation Information

The chapters in this book were originally published in the *British Journal for the History of Philosophy*, volume 23, issue 6 (December 2015). When citing this material, please use the original page numbering for each article, as follows:

Introduction
Monadologies: An Historical Overview
Jeremy Dunham and Pauline Phemister
British Journal for the History of Philosophy, volume 23, issue 6 (December 2015) pp. 1023–1032

Chapter 1
Reconciling Leibnizian Monadology and Kantian Criticism
Richard Mark Fincham
British Journal for the History of Philosophy, volume 23, issue 6 (December 2015) pp. 1033–1055

Chapter 2
Herbart's Monadology
Frederick Beiser
British Journal for the History of Philosophy, volume 23, issue 6 (December 2015) pp. 1056–1073

Chapter 3
Bolzano's Monadology
Peter Simons
British Journal for the History of Philosophy, volume 23, issue 6 (December 2015) pp. 1074–1084

Chapter 4
From Habit to Monads: Félix Ravaisson's Theory of Substance
Jeremy Dunham
British Journal for the History of Philosophy, volume 23, issue 6 (December 2015) pp. 1085–1105

Chapter 5
Reviving Spiritualism with Monads: Francisque Bouillier's Impossible Mission (1839–64)
Delphine Antoine-Mahut

Chapter 6
Learning from Leibniz: Whitehead (and Russell) on Mind, Matter and Monads
Pierfrancesco Basile

Chapter 7
British Idealist Monadologies and the Reality of Time: Hilda Oakeley Against McTaggart, Leibniz, and Others
Emily Thomas

Chapter 8
Heidegger on the Being of Monads: Lessons in Leibniz and in the Practice of Reading the History of Philosophy
Paul Lodge

Chapter 9
Five Figures of Folding: Deleuze on Leibniz's Monadological Metaphysics
Mogens Lærke

Chapter 10
Leibniz's Monadological Positive Aesthetics
Pauline Phemister and Lloyd Strickland

For any permission-related enquiries please visit:
http://www.tandfonline.com/page/help/permissions

Notes on Contributors

Delphine Antoine-Mahut is Professor of Early Modern Philosophy at the École normale supérieure de Lyon, France.

Pierfrancesco Basile is Lecturer in Philosophy at the University of Bern, Switzerland.

Frederick Beiser is Professor of Philosophy at Syracuse University, USA.

Jeremy Dunham is Assistant Professor in Philosophy at the University of Durham, UK.

Richard Mark Fincham is Associate Professor and Chair of Philosophy at the American University in Cairo, Egypt.

Mogens Lærke is Senior Researcher at the Centre national de la recherche scientifique, Paris, France.

Paul Lodge is Professor of Philosophy, based at Mansfield College, University of Oxford, UK.

Pauline Phemister is Professor of the History of Philosophy at the University of Edinburgh, UK.

Peter Simons is Professor Emeritus of Philosophy at Trinity College Dublin, Ireland.

Lloyd Strickland is a Reader in Philosophy at Manchester Metropolitan University, UK.

Emily Thomas is Assistant Professor of Philosophy at the University of Durham, UK.

ARTICLE

INTRODUCTION
MONADOLOGIES: AN HISTORICAL OVERVIEW

Jeremy Dunham and Pauline Phemister

> This introductory overview comprises (i) a brief account of Leibniz's own monadology; (ii) a discussion of the reception of his philosophy up to Kant; and (iii) a short overview of the monadologies developed after Kant's first Critique, made via a summary of key points raised in this guest issue, highlighting recurrent themes, which include questions of historiography.

Gottfried Wilhelm Leibniz's enigmatic *Monadology*, written in 1714, is Leibniz's most succinct and systematic outline of his metaphysics of monads and rightly recognized as a pivotal text in the history of philosophy. It opens with a concise definition of the central notion:

> THE *MONAD*, which we shall discuss here, is nothing but a simple substance that enters into composites – simple, that is, without parts (*Theodicy*, sec. 10).
> (*Monadology*, §1: GP VI 607; AG 213)

As substances, monads are forces. The true nature of substance, Leibniz claims to have discovered in the early 1690s, consists in force (*On the Correction of Metaphysics:* GP IV 469; L 433. See also *New System*: GP IV 472: L 454). All true substances are also indivisible unities, for, as Leibniz said in his 30 April 1687 letter to Arnauld, 'what is not truly *one* entity is not truly one *entity* either' (GP II 97; LA 121). But even though a monad is *one being*, it is not a solitary being. A single indivisible monad can only enter into composite, divided, extended bodies in conjunction with other monads. There exists an infinite plurality of monads that together ground the infinitely divided material world. Without a plurality of monads, there would be no material world.

However, monads do not enter into composite bodies as physical parts. For Leibniz, parts are by definition always homogenous with the whole. The parts of an extended body are smaller bodies, all of which are essentially extended and divided. The division of matter hence proceeds to infinity; every body however small is divided into further extended and divided parts. Monads are indivisible unities and cannot therefore be homogeneous parts of divided aggregate extended things. They are not physical atoms. The role of the monads is quite different, for they are not only indivisible unities, but also indivisible unifiers. Each monad is a metaphysical atom, dominant over a subset of subordinate monads that it unifies to form with itself what Leibniz calls a 'corporeal substance' (To De Volder, 20 June 1703: LV 264–5). Corporeal substances are living, animal-like creatures, comprising the unifying dominant monad – a mind, soul or entelechy – and the subordinate monads that it unites with itself to form the whole living creature. Each subordinate monad is itself dominant over the monads in its own organic body. These dominated monads are in turn dominant over further subordinate monads in their own organic bodies. Thus the universe of monads is a universe of monads whose organic bodies have monads with their organic bodies enfolded within them, to infinity.

A plurality of monads requires that each monad is identifiably distinct from all the others. As indivisible, they cannot be distinguished by their parts, but they can be distinguished by their qualities, namely their perceptions and appetitions. Monads' perceptions are the means by which they each individually represent the world. Monads' appetitions are the actions of the monads' forces that impel them from one perception to the next, ensuring that the monads' perceptions mirror the changes that occur in bodies as they move and resist one another in accordance with the laws of motion or efficient causes. Because each monad represents the whole world, they cannot be distinguished by the content of their perceptions, but they can be identified by the degrees of confusion or distinctness of their perceptions. This in turn is a result of the appetitive force that lies behind each perceptual state, such that the greater the appetition, the more distinct the resultant perception. Hence, ultimately, it is the degree of each monad's active force, its essential nature, that serves to distinguish one monad from another and that determines the particular 'point of view' from which each monad perceives the world (*Principles of Nature and Grace*, §3: GP VI 599; AG 207).

Because they are indivisible and partless, the movement from one perception to the next cannot be accounted for in terms of any internal rearrangement of parts or transference of parts from one monad to another. 'The monads', as Leibniz famously put it, 'have no windows through which something can enter or leave' (*Monadology* §7: GP VI 607; AG 213). Consequently, every change in a monad must already be contained in its essence from the beginning, constituting a 'law of the series' that unfolds its perceptions in sequence and in harmony with the unfolding sequences of perceptions in all other monads. In this way, though there is no actual causal interaction among the monads, the appearance of interaction is maintained by their perfect synchronization:

> this concomitance I maintain is like several different bands of musicians or choirs separately playing their parts, and placed in such a way that they do not see and do not even hear each other, though they nevertheless can agree

perfectly, each following his own notes, so that someone hearing all of them would find a marvelous harmony there. (To Arnauld, 30 April 1687: GP II 95; AG 84)

The 'interconnection or accommodation of all created things to each other' (*Monadology* §56: GP VI 616; AG 220) is one reason why this is the best of all possible worlds. As Leibniz conceives perfection, the way to attain 'as much perfection as possible' is to produce as 'much variety as possible, but with the greatest order possible' (*Monadology* §58: GP VI 616; AG 220). This is achieved through the creation of an infinity of monads, each perceiving the same universe, but each perceiving it from its own individual perspective (*Monadology* §57).

At the time of Leibniz's death in 1716, the high esteem that he had enjoyed for the greater part of his life had all but disappeared. However, his reputation recovered dramatically on the publication of Latin translations of some of his most important works, including the *Monadology*, in the decade following his death. Their publication elicited wide and controversial discussion across Europe. By the middle of the eighteenth century, Leibniz's 'monad' had become a concept of such central importance in German intellectual life that in 1746 Euler wrote that 'the dispute about monads was so lively and general that ... [e]veryone's conversation fell upon monads everywhere and no one spoke of anything else' (Cited and translated in Clark, 'The Death of Metaphysics in Enlightened Prussia', 446). Nonetheless, it was not until the 1760s that the first two collections of his writings that revealed the enormous depth of his thought would be published: Rudolf Erich Raspe's, *Oeuvres philosophiques latines & françoises de feu Mr. de Leibnitz*, which included Leibniz's *New Essays* (published for the very first time), and Louis Dutens's, *Opera Omnia ... nunc primum collecta, in classes distributa, praefationibus et indicibus exornata*.

During the earlier part of the century, Leibniz's philosophy was closely associated with that of Christian Wolff and the latter's philosophy was labelled by the German professor of philosophy and mathematics Georg Bernhard Bilfinger (1693–1750) as 'Leibniz-Wolffian', much to Wolff's disapproval. Despite this association, there were major differences between the two philosophers. For example, Wolff's metaphysics was not an austere metaphysics of monads, but rather a form of dualism. However, Wolff's theory of substance shared a close enough similarity for the concept of the 'monad' to become a significant part of the Leibniz-Wolffian vocabulary, as can be seen in Alexander Gottlieb Baumgarten's famous Leibniz-Wolffian textbook *Metaphysics*, §§392–418.[1] Wolff's greatest debt to Leibniz was for his theory of pre-

---

[1] On the relationship between Wolff and Leibniz's philosophy see Wilson, 'The Reception of Leibniz'; École, 'Wolff était-il leibnizien?'; Lamarra, 'Contexte Génétique et Première Réception de la Monadologie'; Park, 'Le Débat Wolffien Sur L'Idéalisme de Leibniz Lors de la Première Diffusion de la Monadologie Latine'.

established harmony, which Wolff believed was the only satisfactory solution to the problem of the relationship between the mind and the body.

The most historically significant philosophical engagement with Leibniz's thought during the eighteenth century came from Immanuel Kant. Leibniz's influence on Kant ran deep and touched almost every area of his philosophy. Yet at no point in Kant's career could he straightforwardly be regarded as a Leibnizian. Even in his pre-critical days as the author of the *Physical Monadology* (1756) he was profoundly critical of certain characteristic elements of Leibniz's thought. After having published his *Critique of Pure Reason* twenty-five years later, Kant became well known for having provided the most serious and unavoidable challenges for the pursuit of rationalist metaphysics that it had ever encountered. The crucial error of Leibniz's rationalist method, Kant argued, was that Leibniz engaged in logical reflection without recognizing the unique role played by sensibility. He, therefore, '*intellectualised* appearances' (A271/B327). According to Kant, the errors in Leibniz's system are due to his failure to understand that our appearances are conditioned by sensibility and that if a concept is not applied in intuition, then it lacks objectivity. Although such concepts may have a logical meaning, they have no significance in the application to objects. The use of concepts outside of this relation is, Kant writes, 'a mere play of imagination or of understanding' (A239/B298). Therefore, Kant calls his critique of Leibniz 'the critique of pure understanding'. For Kant, '[t]he pure categories, apart from any formal conditions of sensibility, have only transcendental significance; nevertheless they may not be employed transcendentally' (A248/B305). This means that they have transcendental (idealist) significance insofar as they are the necessary conditions for the possibility of experience, but they have no transcendental (realist) use because they cannot be employed in arguments for the existence of entities that take us beyond appearances.

According to the received view, Kant's critical revolution put an end to the kind of metaphysics of which the monadology is the example *par excellence*. This volume will challenge this view and provide a far more nuanced version of philosophy's 'post-Kantian' tradition, spanning from the late eighteenth to the early twentieth century, by bringing to light a rich tradition of new monadologists, many of whom have been unjustifiably forgotten by contemporary historians of philosophy. Through this complex dialogue, the 'monadology' is shown to be a remarkably fecund hypothesis allowing for many possible variations and developments. By focusing on the monadology, therefore, the depth and breadth of the post-Kantian period is exposed in original and previously unexplored ways and the road is laid open for further research.

This guest issue opens with Richard Fincham's 'Reconciling Leibnizian Monadology and Kantian Criticism' in which he discusses some of the first attempts to return to a form of Leibnizian monadology without ignoring the crucial insights of Kant's critical philosophy. Fincham argues that the Polish Lithuanian philosopher Solomon Maimon (1753–1800) and the

German idealist F. W. J. Schelling (1775–1854) both highlight the problematic dualism between the intellect and sensibility at the heart of Kant's philosophy. By focusing on the spirit rather than the letter of Kant's work, Maimon and Schelling attempt to reconcile Leibniz and Kant. While Kant, they argue, cannot explain the connection between the concepts of the understanding and intuitions of sensibility, Leibniz is able to overcome the problem of the two-fold nature of cognitive faculties by conceiving both the understanding and sensibility as arising from the same monadic source.

Maimon's crucial claim is that what Kant conceived as synthetic a priori truths are in fact, insofar as they are in the infinite understanding, analytic truths, and only synthetic truths, insofar as they are appearances in the finite understandings. They are synthetic truths in the latter because of the limitations of the individual's finite perspective. For Leibniz, the monad's limitations give rise to its sense perceptions and serve to differentiate them from each other and from God. The same is true of Maimon's monads, although he expresses the thought in Kantian language as the claim that what is analytic in the infinite appears synthetic in the finite understanding. Fincham shows that Schelling similarly attempts to solve the same Kantian problems through his conception of what he describes as the 'infinite spirit or Absolute subject' that produces finite spirits, which, like monads, are windowless perspectives on the Absolute.

The canonical reconstruction of the history of post-Kantian thought focuses on the monistic aspects that found their ultimate realization in the absolute idealism of Hegel. However, there is a less well-known but just as significant story that emphasizes the Leibnizian monadological pluralist aspects that were being developed even by Hegel's contemporaries, such as Johann Friedrich Herbart (1776–1841) and Bernard Bolzano (1781–1848). Frederick Beiser picks up this later story in his article 'Herbart's Monadology'. While many historians of philosophy have followed Hegel by presenting the history of philosophy leading to him from Kant as a smooth trajectory running through Fichte, Schelling up to Hegel himself, Beiser shows that the truth is much less tidy. While Hegel wanted to trace a history that would be completed in his own absolute Idealism, and as such placed greater emphasis on the monist tendencies in the history of philosophy, Beiser questions this neat sanitized version of the history of post-Kantian thought. Focusing on Herbart, he invokes a pluralist tradition running through the same historical periods, stemming not from Spinoza but from Leibniz. Like Maimon, Herbart develops what he calls a 'critical monadology' that takes seriously Kant's critical philosophy but similarly sees a return to the Leibnizian metaphysics as a solution to the critical system's failures.

Herbart considers his monadology as belonging to the post-rather than pre-Kantian tradition because he agrees with Kant's critique of the rationalist method – especially his denial that existence is a predicate. Whereas for Kant, analysis of experience leads to transcendental idealism, for Herbart, the monadology is discovered through analysis of our experience and its most basic concepts. In keeping with Leibniz, Herbart argues that the rich diverse complexes given in experience must be grounded by simple metaphysical monads, which he calls 'Reals'. Herbart claims that we cognitively construct space and time from our perceptions of simple things. Therefore, in Herbart's account of the relationship between appearance and reality, space and time themselves are appearances of monads, understood in Leibnizian fashion as simple and self-preserving beings.

Peter Simons shows that Bolzano's monadology develops the Leibnizian metaphysics in an even more realist fashion. Bolzano's is a physical monadology, not unlike that of the early Kant. Bolzano's monads are simple, fundamental, yet they exist in space. Bolzano represents a crucial chapter in the history of monadological metaphysics insofar as he made one of the boldest attempts to construct a *physical* monadology that could be reconciled with classical physics whilst using the rationalist a priori methodology, even though, as Simons argues here, the attempt was ultimately unsuccessful. The rationalist approach to the monadology is shown to fail, he argues, when the monads are understood purely as physical atoms.

In agreement with Simons, the French monadologists, considered here in papers by Jeremy Dunham and Delphine Antoine-Mahut, believed that a consistent monadology could not be developed using the rationalist a priori method. Nonetheless, the French monadologists argued that Leibniz himself combined both a priori reasoning, on the one hand, with empirical evidence gleaned from introspective examination of the activities of one's own mind, on the other. However, they maintained that the conclusions Leibniz drew when using the a priori method took him down a path that would, if continued, lead to Spinozistic pantheism and its consequent threats to freedom and individuality, while conversely the conclusions that he drew from the empirical side proved the existence of free individuals. Faced with these opposing and irreconcilable consequences of rationalist and empiricist, methodologies, the French monadologists chose to promote the latter. In contrast to the earlier thinkers from the Germanic tradition, they considered Kant's critical philosophy as a dangerous scepticism rather than a profound philosophical revolution. Controversially, in 'From Habit to Monads: Félix Ravaisson's Theory of Substance', Dunham argues that reading Ravaisson's (1813–1900) philosophy as a dialogue with the post-Kantian tradition has been one of the major sources for misinterpretation of his work. Instead, Dunham shows that, correctly understood, Ravaisson uses the analysis of habit as an attempt to argue from empirical reflection to a pluralist metaphysics. In Antoine-Mahut's paper, 'Reviving Spiritualism with Monads: Francisque Boullier's Impossible Mission (1839–64)', we see how this French tradition continued to develop throughout the nineteenth century and why Francisque Bouillier (1813–1899) believed that the monadology could be used to provide an alternative both to the passive Cartesian mechanism and to Scholastic animism and therefore to provide a metaphysics genuinely compatible with developments in the life sciences.

Antoine-Mahut makes the broad historiographical point that the way that Leibniz was understood and interpreted in this period was never as a philosopher in isolation but always in constellation both with his contemporaries Descartes and Stahl and also with the dialogues being developed within the French context itself. While attempts were made to understand Leibniz in his own intellectual context, for example, in relation to Descartes and Stahl, his views were also regarded as making a significant contribution to the contemporary philosophical scene in nineteenth-century France. Politically, it was advantageous to be able to align one's own philosophy with that of the great Leibniz. Historiographically, of course, recognizing that one's own reading of the history is itself historically situated calls into question the very possibility and even desirability of reaching absolute objective truths in matters historical. By reflecting on this historical treatment of the history of philosophy Antoine-Mahut questions whether objectivity is something that could ever be obtained.

In Britain too, Leibniz's philosophy came to be seen not as an historical relic but as a living system to be worked with and adapted to suit a modern context. Perhaps the most ambitious attempt to develop the theory of monads into a full-blown metaphysical system compatible with the revolutions in both physics and biology that had occurred since Leibniz came from Alfred North Whitehead (1861–1947). As Pierfrancesco Basile shows in his article 'Learning from Leibniz: Whitehead (and Russell) on Mind, Matter, and Monads', Whitehead took issue with Leibniz's substance ontology, subject-predicate logic and with his system of pre-established harmony. In its place, Whitehead develops an ontology that sees the building blocks of reality as experiential processual units, called 'actual occasions'. Whitehead himself admits that his theory of actual occasions is a theory of monads, but Whitehead's monads progress through their mutual creative interaction rather than unfolding from their own substantial essences as do Leibniz's. Although Basile argues that Whitehead's attempt to reconstruct this theory in this way was not entirely successful, he nonetheless claims that Whitehead's philosophy of organism is importantly suggestive for reconsidering the relationship between mind and matter.

Whitehead was not the only British philosopher to construct a theory of monads during the first half of the twentieth century. Herbert Wildon Carr (1857–1931), James Ward (1843–1925), John McTaggart Ellis McTaggart (1866–1925) and Hilda Oakeley (1867–1950) all developed original monadologies. In her paper 'British Idealist Monadologies and the Reality of Time', Emily Thomas focuses on McTaggart and Oakeley and their attempts to understand the nature of time within a monadological framework. McTaggart's arguments for the rejection of the reality of time are well-known, but Oakeley's critique of McTaggart and ultimate defence of the reality of time are much less so. Oakeley's main argument again starts from experience and our personal perceptions of temporal passage. Thomas defends Oakeley's argument and she also shows that Oakeley's argument is generalizable to all monadologies. Any consistent monadological metaphysical system, Thomas contends, must affirm the existence of time.

What we see in all these philosophers are attempts to engage with the history of philosophy in a way that Paul Lodge calls 'dialogical history'. That is to say, each of them develops their own philosophical position in dialogue or through dialogue with Leibniz and his texts. In his paper 'Heidegger on the Being of Monads', after discussing Martin Heidegger's (1889–1976) own attempt at dialogical history with Leibniz in his *Metaphysical Foundations of Logic*, Lodge defends the virtues of a methodological pluralism in the history of philosophy. The philosophical enterprise needs not only 'exegetical' historians of philosophy, but historical philosophers, and dialogical and creative historians of philosophy too, for, as Lodge argues, it is only when all four types of philosopher work together in a mutually collaborative intellectual enterprise that the best use is made of our discipline's rich history. Furthermore, Lodge's account of Heidegger's engagement with Leibniz uncovers a number of non-standard and interesting aspects of Heidegger's reading of the *Monadology* from which Leibniz scholars can learn. These include his understanding of the unifying nature of monads, force understood as *drive*, and the latter's relationship with perception and appetite. Like a number of the monadologists discussed in this volume, Heidegger emphasizes the importance of Leibniz's use of introspection for understanding the nature of monads. Importantly, Lodge shows how this focus on introspective methodology can be used to provide Leibniz scholars with hints on how to interpret Leibniz's own understanding of monads.

What Mogens Lærke finds in Gilles Deleuze's (1925–1995) relationship with Leibniz is perhaps most aptly situated within the category of 'creative history of philosophy'. According to Lærke, Deleuze's *Le Pli* is not primarily a reading of Leibniz; it is instead a work of aesthetics that draws on an ahistorical reading of Leibniz's thought as a 'monadological metaphysics'. In his paper, 'Five Figures of Folding. Deleuze on Leibniz's monadological metaphysics', Lærke offers an exceptionally clear and insightful exposition of Deleuze's use of Leibniz's notion of the folds of bodies, of events, and of the monads that provide their logical and existential grounding. Although Lærke provides evidence that Deleuze's focus on the fold is textually better grounded than the latter could have realized, he is critical of Deleuze's 'stubbornly synchronic' approach to reading Leibniz. Therefore, he does not provide a defence of Deleuze's reading, but rather he shows through the use of graphic figures that Deleuze's use of the 'fold-concept' captures many of the basic elements of Leibniz's philosophy in an 'extraordinarily synthetic way'. Importantly, by doing so Lærke provides the key for Leibniz scholars to understand the specific contribution that Deleuze's work makes to the scholarship, and at the same time the key for Deleuzians to understand the core insights that Deleuze takes from Leibniz's work.

Leibniz's own aesthetics is the theme of the closing paper in this volume. Leibniz composed no treatise on aesthetics and his writings only rarely address the topic of beauty *per se*. However, his definition of beauty as 'that, the contemplation of which is pleasant' (*Elements of Natural Law*: A VI i 464: L 137) and his account of the perfection of individuals and of the world in terms of the maximization of ordered variety gave rise to Baumgarten's rationalist aesthetics in the eighteenth century. In the context of contemporary positive aesthetics, it remains highly relevant today. In 'Leibniz's Monadological Positive Aesthetics', Pauline Phemister and Lloyd Strickland argue that the objective beauty of Leibniz's monadological world, discoverable by natural scientific investigation but ultimately justified theologically by appeal to God's decision to create the best possible world, constitutes a positive aesthetics that is both clearer and stronger than the secular positive aesthetics advanced in recent years by Allen Carlson. The rich history of the monadology shows that it has – for the last 300 years – been regarded as a living philosophy, and that it is a remarkably fecund philosophical hypothesis capable of many possible developments without losing its distinctively Leibnizian character. Far from being extinguished by the Kantian critical revolution, therefore, the monadology continues to have relevance today. Phemister and Strickland focus on just one area of many in philosophy where it may have a future too.[2]

BIBLIOGRAPHY

A = Leibniz, Gottfried Wilhelm. *Sämtliche Schriften und Briefe*. Edited by Berlin-Brandenburgische Akademie der Wissenschaften. Berlin: Akademie Verlag, 1923–.

AG = Leibniz, Gottfried Wilhelm. *G. W. Leibniz: Philosophical Essays*. Edited and Translated by Roger Ariew and Daniel Garber. Indianapolis: Hackett, 1989.

Baumgarten, Alexander Gottleib. *Alexander Baumgarten, Metaphysics: A Critical Translation with Kant's Elucidations*. Edited and translated by Courtney D. Fugate and John Hymers. London: Bloomsbury Academic, 2013.

---

[2] We gratefully acknowledge the support of the Leverhulme Trust, the Institute for Advanced Studies in the Humanities, and the British Society for the History of Philosophy. We would also like to thank the Editor of this Journal, Mike Beaney, for his support and advice and Jessica Leech for commenting on a draft of this introduction. We extend our thanks also to each of the contributors.

Clark, William. 'The Death of Metaphysics in Enlightened Prussia'. In *The Sciences in Enlightened Europe*, edited by William Clark, Jan Golinski, and Simon Schaffer, 423–76. Chicago: University of Chicago Press, 1999.

Dutens, Ludovici, ed. *Opera Omnia ... nunc primum collecta, in classes distributa, praefationibus et indicibus exornata*. 6 vols. in 7. Geneva: apud fratres de Tournes, 1768.

École, Jean. 'Wolff était-il leibnizien?' In *Nouvelles études et nouveaux documents photographiques sur Wolff*, edited by Jean École, 131–51. Hildesheim: Georg Olms, 1997.

GP = Leibniz, Gottfried Wilhelm. *Die Philosophischen Schriften von Gottfried Wilhelm Leibniz*. Edited by C. I. Gerhardt, 7 vols. Berlin: Weidman. Reprint, Hildesheim: Olms, 1965, 1875–90.

Kant, Immanuel. *The Critique of Pure Reason*. Translated by Norman Kemp Smith. London: Macmillan, 1933.

LA = Leibniz, Gottfried Wilhelm. *The Leibniz-Arnauld Correspondence*. Edited and translated by Haydn Trevor Mason. Manchester: Manchester University Press, 1967.

Lamarra, Antonio. 'Contexte Génétique et Première Réception de la Monadologie: Leibniz, Wolff et la Doctrine de L'Harmonie Préétablie'. *Revue de synthèse* 128, nos 3–4 (2007): 311–23.

LV = Leibniz, Gottfried Wilhelm. *The Leibniz-De Volder Correspondence*. Edited and translated by Paul Lodge. New Haven and London: Yale University Press, 2013.

Park, Jeongwoo. 'Le Débat Wolffien Sur L'Idéalisme de Leibniz Lors de la Première Diffusion de la Monadologie Latine'. *Revue de synthèse* 128, nos. 3–4 (2007): 325–39.

Raspe, Rudolf Erich. *Oeuvres philosophiques latines & françoises de feu Mr. de Leibnitz*. Amsterdam and Leipzig: Chez Jean Schreuder, 1765.

Wilson, Catherine. 'The Reception of Leibniz'. In *The Cambridge Companion to Leibniz*, edited by Nicholas Jolley, 442–74. Cambridge: Cambridge University Press, 1995.

ARTICLE

RECONCILING LEIBNIZIAN MONADOLOGY AND KANTIAN CRITICISM

Richard Mark Fincham

This paper (written for the '300 Year's of Leibniz's Monadology' conference) explores systematic parallels between the criticisms of Kantian cognitive dualism provided by Salomon Maimon within his 'Essay on Transcendental Philosophy' of 1790 and F.W.J. Schelling within his 'General Overview of the Most Recent Philosophical Literature' of 1797. It discusses how both Maimon and Schelling suggest that the difficulties with Kant's cognitive dualism are so severe that they can only be resolved by recourse to a Leibnizian position, in which sensibility and understanding, and matter and form, arise from one and the same cognitive source. It thus shows how Maimon and Schelling – within 1790 and 1797, respectively – sketch systems of transcendental philosophy explicitly modelled on the Leibnizian philosophy, which both of them interpret as claiming that God is immanently contained within the human soul.

In 1790 the *Prussian Royal Academy of Sciences* announced a prize essay contest with the question, *What Real Progress Has Metaphysics Made in Germany since the Time of Leibniz and Wolff?* Fragments of Immanuel Kant's unfinished response remain. Therein Kant interprets Leibniz as a recent exponent of that theoretico-dogmatic metaphysics which constitutes one pole of the antinomical conflict in which reason reaches a 'sceptical standstill' but which his own 'criticism' has resolved (see AA vol. 20, 277–292).[1] Indeed, Kant described the thesis of the 'Second Antinomy' as 'the dialectical principle of *monadology*' (A442/B470). Accordingly he

[1] Kant's works are cited according to the Akademie Ausgabe (AA) (i.e. Kants, *Gesammelte Schriften*) pagination with the exception of the *Critique of Pure Reason* for which the standard A/B pagination is provided. In all cases the Kant, *Cambridge Edition* translations have been followed.

sees Leibnizianism as an illusory metaphysics thoroughly undermined by the *Critique of Pure Reason* and replaced by his own true metaphysics. This interpretation proved influential, yet was not shared by many of Kant's heirs and successors.

In 1797 Friedrich Wilhelm Joseph Schelling's 'General Overview of the Most Recent Philosophical Literature'[2] considered the Academy's question and criticized the response of one of the joint winners, the Leibnizian, Johann Christoph Schwab. Given his reputation as an Idealist successor of Kant with Spinozistic proclivities, it might surprise us to discover that Schelling criticizes Schwab for failing to laud Leibniz's continuing relevance. 'The *Leibnizian* philosophy' Schelling writes 'has, perhaps more than any other, its spirit and its letter [*ihren Geist und ihren Buchstaben*]' and has 'suffered' from both its detractors and supporters focusing upon the latter to the detriment of the former (Schelling, *Historisch-kritische Ausgabe*, vol. I, 4, 96–97). A focus upon its 'spirit' reveals that 'in regard to the fertility of its ideas' the Leibnizian system is 'the only one of its kind that is capable of a truly infinite development' (Schelling, *Historisch-kritische Ausgabe*, vol. I, 4, 99). Schelling justifies this judgement with reference to Leibniz's claim to have found a '"*perspectival middle-point*" from where regularity and agreement can be seen within the chaos of multifarious opinions' (Schelling, *Historisch-kritische Ausgabe*, vol. I, 4, 98).[3] He thus credits Leibniz with a view that he himself endorses, namely that we should have before our eyes the 'idea' of *one* true philosophical system – an archetype to which all products of genuine 'philosophical talent' approximate (Schelling, *Historisch-kritische Ausgabe*, vol. I, 4, 97–98). Accordingly, if one possesses 'philosophical talent' and thus focuses upon the 'spirit' of a genuine philosophical system, one will see it as a member of that one universal system in which all products of philosophical talent are organically united. In this way, Schelling suggests that, far from having been supplanted by Kantian criticism, the Leibnizian system should be reconcilable with Kantianism (Schelling, *Historisch-kritische Ausgabe*, vol. I, 4, 94). Indeed, Schelling also tells us that 'in addition to the *literal language* [*Wort-Sprache*]' of Kant's works 'there also exists a *language of spirit* [*Sprache der Geister*]' and that Kant has been 'misunderstood' by both his supporters and opponents because they focused upon the former to the detriment of the latter (Schelling, *Historisch-kritische Ausgabe*, vol. I, 4, 69; Pfau, *Idealism and the Endgame of Theory*, 68). Whereas Kant

---

[2]This was originally published in instalments in the *Philosophisches Journal einer Gesellschaft Teutscher Gelehrten* between February 1797 and November 1798. A slightly abridged and amended version reappeared within Schelling's *Philosophische Schriften* of 1809 with the title *Abhandlungen zur Erläuterung des Idealismus der Wissenschaftslehre*. The second version of the text is translated in Pfau, *Idealism and the Endgame of Theory*, 62–138.

[3]Schelling is here referencing a passage from Leibniz's *Recueil de div Pieces par des Maizeaux* (see Schelling, *Historisch-kritische Ausgabe*, vol. I, 4, 327).

had, for the benefit of speculation, separated human knowledge into distinctive components, the lack of 'philosophical talent' of both his supporters and opponents, means they are incapable of reuniting what has been separated, and realizing that what has been separated can never be separated in reality (Schelling, *Historisch-kritische Ausgabe*, vol. I, 4, 77–78; Pfau, *Idealism and the Endgame of Theory*, 73–74). If one possesses 'philosophical talent' and thus focuses upon the 'spirit' of Kant's work, then, Schelling somewhat controversially suggests, one would reunite that which was separated, and realize that Kant in no way espouses a cognitive dualism in which there is 'an utter separation' of understanding and sensibility, ideality and reality, and form and matter, but rather claims that they are in reality one (Schelling, *Historisch-kritische Ausgabe*, vol. I, 4, 77 & 91; Pfau, *Idealism and the Endgame of Theory*, 73 & 82–83). Schelling thus interprets Kant as claiming that 'both matter and form must primordially emerge and originate from *within ourselves*' (Schelling, *Historisch-kritische Ausgabe*, vol. I, 4, 91; Pfau, *Idealism and the Endgame of Theory*, 83).

To account for Leibniz's change of fortune, this article will focus upon the work of another contributor to the Academy's contest, Salomon Maimon. Maimon's contribution, 'Concerning the Progress of Philosophy', claims to 'reconcile Leibniz with the *Critique of Pure Reason*' (Maimon, *Gesammelte Werke*, vol. 4, 51). This is possible, Maimon contends, since, if correctly understood, Leibniz's monadology is, *contra* Kant, not a metaphysical doctrine resolving the world into an infinity of simple things in themselves, but rather – like Bonaventura Cavalieri's method of indivisibles within mathematics – a doctrine of '*fictions*', *useful* for explaining cognition of nature (Maimon, *Gesammelte Werke*, vol. 4, 51–54 & 78). This article shows how Maimon drew upon this reading of Leibniz within his earlier *Essay on Transcendental Philosophy* of 1790, a work which (like Schelling's 'General Overview') argues that Kant's (supposed) cognitive dualism is so problematic that it requires Leibnizian resources to be salvaged, to thus offer 'a *coalition-system*' appropriating both Kantian and Leibnizian insights (Maimon, *Gesammelte Werke*, vol. 1, 557). We shall begin by (1) discussing the problems Maimon locates in Kant's work and then (2) show how the attempt to resolve these difficulties leads him to embrace a Leibnizian position. Finally, (3) we shall see how many aspects of Maimon's resolutions to these problems are appropriated by Schelling in 1797.

1. KANTIAN PROBLEMS

Maimon's reading of Kant very much focuses upon his 'literal language'. According to Maimon, Kant's work rests on a fundamental assumption concerning our relation to the world. Thus, whereas Hume starts from the assumption that we are ultimately only conscious of a subjective sequence of perceptions, Kant starts from the assumption that everyday consciousness is related to both a subjective sequence of perceptions within us and an objective sequence of events in space. The latter consciousness finds expression within the truth-claims of natural science that describe necessary connections within the world and claim objective validity (see B141–142 & AA vol. 4, 298–301). For Maimon, Kant assumes from the outset that the natural scientist is justified in issuing such judgements. He thus writes that 'Kant presupposes as indubitable the fact that we possess propositions of experience (expressing necessity)' (Maimon, *Gesammelte Werke*, vol. 2, 186; Maimon, *Essay on Transcendental Philosophy*, 100). And since consciousness of necessarily connected states of affairs constitutes experience (*Erfahrung*) in Kant's sense, Maimon says Kant simply presupposes that such experience is 'actual' (Maimon, *Gesammelte Werke*, vol. 2, 186; Maimon, *Essay on Transcendental Philosophy*, 100) Accordingly, Maimon bemoans the paucity of Kant's response to the question *quid facti?*, the question concerning whether we actually do possess such experience at all (Maimon, *Gesammelte Werke*, vol. 2, 70–71; Maimon, *Essay on Transcendental Philosophy*, 42).

Presupposing that such experience is actual, much of Kant's speculative philosophy is concerned with how it is possible. His answer is that it is conditioned by *a priori* concepts of the understanding or categories, under which given intuitions are subsumed. He thus answers Hume's claim that the idea of causality arises from our imagination compelling us to believe that constantly conjoined perceptions are necessarily connected, since, according to Kant, causality is one of those *a priori* categories conditioning experience, and thus it has an objective as opposed to subjective validity (AA vol. 4, 312–313). Kant therefore – according to Maimon – proves the 'objective reality' of the categories by showing that they make possible that fact of experience which he presupposed as indubitably actual (Maimon, *Gesammelte Werke*, vol. 2, 186; Maimon, *Essay on Transcendental Philosophy*, 100). But while Kant's attention to the question *quid facti?* may be negligible, the same cannot be said about his attention to the question *quid juris?*, the question concerning the *right* with which *a priori* categories are applied to *a posteriori* intuitions to make experience possible. According to Maimon, Kant correctly recognizes that it is impossible for *a priori* categories to be legitimately applied directly to anything given *a posteriori*, and thus attempts to answer the question *quid juris?* by identifying something *a priori* within empirical intuitions that can bridge the gap between *a priori* categories and the *a posteriori* given (Maimon, *Gesammelte Werke*, vol. 2, 363; Maimon, *Essay on Transcendental Philosophy*, 187). Kant thus identifies time as a form of all intuition, which, insofar as it logically precedes the *a posteriori* data organized within it, is, like the categories, *a priori*. For Kant, therefore, we do not directly apply categories to the *a posteriori* content of empirical intuition, but rather to a particular *a priori* determination (i.e. transcendental schema) of the *a priori* form of intuition (i.e. time) in which the *a posteriori* content of any intuition is organized.

Maimon's divergence from Kant stems from his dissatisfaction with his response to both the question *quid facti?* and the question *quid juris?* In response to the question *quid facti?*, Maimon suggests that Kant's presupposition that the fact of experience is indubitable cannot silence Hume's doubts. Referencing Kant's example of a 'judgement of experience', that 'the fire warms (makes warm) the stone', Maimon notes that, for Hume, this proposition only describes an association of ideas that has become subjectively necessary as a result of constant conjunction, and not objectively valid insight into necessary connections in the world – and that the same thing can be said of any 'judgement of experience' (Maimon, *Gesammelte Werke*, vol. 2, 72–74; Maimon, *Essay on Transcendental Philosophy*, 42–43). Conversely, Maimon argues that, even if we assume that we have some consciousness of objectively valid states of affairs, we would still need experience of constant conjunction to ascertain which objective states of affairs are necessarily connected. For example, if both (*b*) warm air and (*c*) a snowfall are perceived soon after (*a*) the lighting of a stove, we do not regard *b* but not *c* as a necessary consequence of *a* because *c* unlike *b* is a merely subjective state of affairs. It would therefore seem that our conviction that *a* causes *b* but not *c* can only be explained in terms of our previous experience of having found *a* and *b* – but not *a* and *c* – to be constantly conjoined. But to concede this is to grant Hume's claim. And if the category of causality has no objective reality in any such particular cases, then 'it would also fail to have objective reality in its use in general, because we would not in fact possess any objective sequences' (Maimon, *Gesammelte Werke*, vol. 2, 372; Maimon, *Essay on Transcendental Philosophy*, 191). Maimon therefore sees no reason for why Kant's assumption that consciousness is related to both a subjective sequence of perceptions and an objective sequence of events should be privileged over Hume's competing assumption that we are conscious of only subjective sequences of perceptions. Whereas Kant saw the *fact* of experience as indubitable, Maimon believes that it may be denied.

This sceptical response to the question *quid facti?* presents a serious challenge. For, Kant does indeed frequently begin arguments with the fact of experience, before separating its distinctive components (e.g. A22/B36). But if we doubt the premises we must doubt the conclusions. Hence we may doubt whether space and time are pure forms of intuition logically preceding the given material content of experience, and also whether the categories possess objective validity. Maimon thus tells us that doubting the fact of experience means that 'we would have nothing solid to lean on in determining the reality of the categories' (Maimon, *Gesammelte Werke*, vol. 2, 71; Maimon, *Essay on Transcendental Philosophy*, 42) and thus that 'I cannot prove their objective validity the way Kant does' (Maimon, *Gesammelte Werke*, vol. 2, 186; Maimon, *Essay on Transcendental Philosophy*, 100). However, even if we grant Kant's claims concerning experience,

Maimon argues that we should still reject Kant's claims concerning the objective reality of the categories, since he also believes that Kant has inadequately answered the question *quid juris?* – and thus has not provided a completely comprehensible account of how the categories are applicable to intuitions. As we saw, Kant attempts to explain how *a priori* categories could be applied to the empirical content of intuition by appealing to something which these two heterogeneous elements have in common, namely, temporal determinations. Maimon objects, however, that this supposed solution immediately generates another version of the problem it was intended to solve. For Kant insists that sensibility and understanding are completely heterogeneous faculties, thus meaning that a connection between *a priori* categories and *a priori* forms of intuition is, just as much as any connection between *a priori* categories and the *a posteriori* given, a connection between two completely heterogeneous elements.

Confronted with Maimon's doubts concerning the fact of experience, we might be surprised to subsequently learn that he is still interested in providing a satisfactory answer to the question *quid juris?* Maimon is aware, however, that, according to Kant, one must appeal to the supposed connection between understanding and intuitions to explain the possibility of *a priori* synthetic cognition within mathematics. Maimon agrees with Kant that mathematical propositions are apodictic and also that they are – at least for us – synthetic, but he disagrees with Kant's account of the possibility of such cognition. Maimon distinguishes two senses in which something's possibility can be explained. An explanation of something's possibility can be either 'the explanation of the meaning of a rule of condition, i.e. the demand that a merely symbolic concept be made intuitive' or 'the genetic explanation of a concept whose meaning is already familiar to us' (Maimon, *Gesammelte Werke*, vol. 2, 58; Maimon, *Essay on Transcendental Philosophy*, 35). Maimon believes that Kant's account of the possibility of mathematical propositions does not succeed in explaining their possibility in the second sense, but merely explains their possibility in the first sense, i.e. in appealing to the fact that mathematical concepts can be 'constructed' within intuition. Such an explanation, however, will only satisfy someone convinced by Kant's account of space and time as *a priori* forms of intuition which specifically *precede* material content. Anyone unconvinced by this specific claim about intuitions, like Maimon himself, will, in contrast, be unable to appeal to these forms of intuition to explain why any mathematical proposition possesses objective as opposed to subjective necessity. Maimon illustrates this with regard to Kant's example that 'the straight line between two points is always the shortest' to suggest that the fact that it can continually be exhibited within intuition in no way explains why this proposition possesses apodictic certainty and objective validity as opposed to a high degree of probability (see Maimon, *Gesammelte Werke*, vol. 2, 364; Maimon, *Essay on Transcendental Philosophy*, 187). To satisfactorily explain this we would have to appeal to the second sense of the explanation of possibility, and explain why the concept of a straight line arises in such a way that the concept of shortness is necessarily connected with it. Maimon thus says that '*Quid juris?* for me means the same as *quid rationis?*' (Maimon, *Gesammelte Werke*, vol. 2, 364; Maimon, *Essay on Transcendental Philosophy*, 187). For if we are to explain *why* the understanding's concept of a straight line is universally found in intuition to be shortest then we need to make comprehensible *how* the very concept of a straight line is necessarily connected with the predicate of shortness. Maimon is convinced, however, that the only way that such a proposition could possess both apodictic certainty and objective validity is if it were based upon the principle of non-contradiction and therefore analytic – even while, at the very same time, agreeing with Kant that, for us, such propositions are synthetic. Maimon therefore finds himself in the position of being convinced that mathematical truths possess apodictic certainty and, hence, objective validity with respect to our intuitions, while nevertheless being convinced that any apodictically certain, objectively valid truth can only be based upon the principle of non-contradiction; and yet, at one and the same time, even if some acceptable definitions were to render them analytic, he is still convinced that the bulk of our mathematical cognitions are, for us, synthetic. Maimon's positive innovations in the sphere of transcendental philosophy arise as a result of his attempts to resolve this apparent paradox.

In analogy with Kant's supposed procedure of presupposing the fact of experience and then asking how it is possible, Maimon suggests that his own philosophical project arises as a result of starting out from the indubitability of mathematics, and then asking how it can be made comprehensible. He writes:

> I also take a fact as ground, but not a fact relating to *a posteriori* objects (because I doubt the latter) but a fact relating to *a priori* objects (of pure mathematics) where we connect forms (relations) with intuitions; and because this undoubted fact refers to *a priori* objects, it is certainly possible, and at the same time actual. But my question is: how is it comprehensible? (*Quid juris?* for me means the same as *quid rationis?* because what is justified is what is legitimate, and with respect to thought, something is justified if it conforms to the laws of thought or reason.) *Kant* shows merely the *possibility* of his fact, which he merely *presupposes*. By contrast, my fact is *certain* and also *possible*. I merely ask: what sort of hypothesis must I adopt for it to be *comprehensible*? So my question does not belong to transcendental philosophy. But because my solution is universal, and can accordingly be used in relation to the objects of transcendental philosophy ... I therefore think I am justified in introducing it here.
> (Maimon, *Gesammelte Werke*, vol. 2, 363–364; Maimon, *Essay on Transcendental Philosophy*, 187)

The hypothesis which Maimon adopts to explain how mathematical propositions can, for us, be both apodictic and synthetic, while, at the same time, any apodictically and objectively valid proposition must be analytic, claims that, for an 'infinite understanding' which thinks all possible concepts, all apodictic and objectively valid propositions are analytic, while nonetheless appearing synthetic to us on account of our finitude. Accordingly, the fact that 'the straight line between two points is always the shortest' is *comprehensible* on account of it being grounded by the infinite understanding's concept of a straight line, in which the predicate of shortness is necessarily contained (see Maimon, *Gesammelte Werke*, vol. 2, 181; Maimon, *Essay on Transcendental Philosophy*, 97); for us, meanwhile, this proposition remains synthetic on account of the fact that, due to our finitude, we cannot have a complete insight into the concepts thought by an infinite understanding. It is precisely this hypothesis that Maimon believes can be employed within transcendental philosophy to explain the connection between *a priori* concepts and empirical intuitions. For even though Maimon doubts that we possess 'experience' in Kant's sense, he still believes that we require an explanation of the possibility of the 'experience' that we do possess, since he believes that even if our experience is constituted by nothing other than a subjective sequence of perceptions, *a priori* categories in fact serve as conditions of possibility of perception itself (see Maimon, *Gesammelte Werke*, vol. 2, 214–215; Maimon, *Essay on Transcendental Philosophy*, 114).

2. MAIMON'S LEIBNIZIAN SOLUTIONS

To overcome the difficulties within Kantian transcendental philosophy, Maimon promulgates a position which he describes as both rational dogmatism and empirical scepticism (Maimon, *Gesammelte Werke*, vol. 2, 430–432; Maimon, *Essay on Transcendental Philosophy*, 114). According to his system, there is only one *a priori* form of thought that has objective validity, namely, the principle of non-contradiction. All other *a priori* forms are declared to possess only subjective reality, even though their universality means that they 'serve in exactly the same way as if they had objective reality' (Maimon, *Gesammelte Werke*, vol. 2, 432–433; Maimon, *Essay on Transcendental Philosophy*, 222). Accordingly, Maimon proposes a philosophical system in which an infinite understanding thinks 'objects of reason' of which we can only attain 'confused representations' within intuition (Maimon, *Gesammelte Werke*, vol. 2, 432; Maimon, *Essay on Transcendental Philosophy*, 222). For the infinite understanding all truths are analytic, although, because of our limitations, the very same truths will frequently appear to us as synthetic. Maimon thus writes that: 'for the infinite understanding [our] assertoric-synthetic propositions must be apodictic and the apodictic-synthetic propositions analytic' (Maimon, *Gesammelte Werke*, vol. 2, 93; Maimon, *Essay on Transcendental Philosophy*, 53). In this regard Maimon understands himself as following in Leibniz's footsteps.

Indeed, other than himself, Leibniz would be the only philosopher likewise deserving the title of 'rational dogmatist'. He writes: 'If I am asked: who are these rational dogmatists? then for the moment I can name no one but myself. But I believe that this is the Leibnizian system (if it is understood correctly)' (Maimon, *Gesammelte Werke*, vol. 2, 433; Maimon, *Essay on Transcendental Philosophy*, 222). This is the second time within the *Essay* that Maimon suggests that parallels between his own position and that of Leibniz will only be apparent to those who have correctly understood the latter. And it may well be Kant who is being implicitly reproached for failing to grasp the 'spirit' of the Leibnizian system here. Before its publication, after reading the first couple of chapters of the *Essay*, Kant expressed misgivings about its interpretation of Leibniz that were communicated to Maimon:

> [Maimon] seeks ... to demonstrate that *a priori* synthesis can possess objective validity only because the divine understanding, of which our understanding is only a part ... is itself the originator of the forms and of the possibility of the things in the world (in themselves). But I very much doubt that this was Leibniz or Wolff's meaning.
> (AA vol. 11, 49–50; Maimon, *Essay on Transcendental Philosophy*, 231)

In the two passages in which Maimon suggests that one must 'correctly understand' Leibniz in order to realize his proximity to him, both of which were subsequently appended to the manuscript which Kant read, Maimon seemingly retorts that it is he himself, rather than Kant, who has in fact correctly grasped the 'spirit' of Leibniz. Accordingly, we may assume that Maimon does read Leibniz as claiming (1) that our understanding is a part of God's understanding[4] and (2) that the forms and possibility of things in the world originate with the latter.

---

[4]On occasions Maimon explicitly casts doubt upon the idea that Leibniz himself conceived the human soul as possessing a substantiality separable from God's (e.g. Maimon, *Gesammelte Werke*, vol. 4, 414). Nonetheless, this claim that our understanding is a 'limitation' of the divine understanding seems to have closer affinities with Spinoza. Yitzhak Y. Melamed argues that Maimon was primarily a Spinozist 'cautiously' masquerading as a Leibnizian. While it cannot be denied that Spinoza *also* influenced Maimon, it is possible that Melamed somewhat overstates his case. Indeed, it must be observed that, even while making the Spinozistic claim that the human understanding is a part of God's understanding, Maimon nonetheless, unlike Spinoza, remains an idealist, suggesting, like Leibniz, a thorough reduction of all things to thought (Melamed, 'Salomon Maimon and the Rise of Spinozism', 75). Even if, as Melamed demonstrates ('Salomon Maimon and the Rise of Spinozism', 76–77), idealist misconstruals of Spinoza were prevalent within the eighteenth century, the fact that Maimon does not attribute extension to God may make us question the extent of Spinoza's influence upon him. Furthermore, Melamed admits that 'Maimon did adopt some important doctrines from Leibniz' ('Salomon Maimon and the Rise of Spinozism', 93) and as an example refers to the theory of differentials introduced to resolve the *quid juris?* problem within the *Essay*. The truth of the matter is most probably that which is suggested by Maimon's description of the latter as containing a '*coalition-system*' (Maimon, *Gesammelte Werke*, vol. 1, 557) appropriating the insights of Hume, Kant, Spinoza and Leibniz.

To understand Maimon's reading of Leibniz, it is useful to consider some of Kant's remarks in the 'B Deduction'. When wrestling with the question *quid juris?*, Kant frequently draws attention to his own doctrine of cognitive dualism by contrasting the twofold nature of the human cognitive faculties with those of a divine being whose 'understanding ... itself intuited' (B145). Such a 'divine understanding', possessing intellectual intuition, would constitute a being that 'would not represent given objects, but through whose representation the objects would themselves at the same time be given, or produced' (B145),[5] and would thus be 'an understanding through whose representation the objects of this representation would at the same time exist' (B139). Within the divine understanding there is therefore no distinction between representation and thing *and* possibility and actuality. Since, for Maimon, the heterogeneity of sensibility and understanding rendered the question *quid juris?* 'insoluble' (Maimon, *Gesammelte Werke*, vol. 2, 63; Maimon, *Essay on Transcendental Philosophy*, 38), he is clearly inspired to view the (for Kant) diametrically opposed faculty of 'an infinite understanding' which 'represents everything to itself intuitively [*intuitive*]' (Maimon, *Gesammelte Werke*, vol. 2, 65; Maimon, *Essay on Transcendental Philosophy*, 38) as providing a clue to a satisfactory answer to that question. 'Difficulties of this kind', he thus writes, could be 'overcome' on the assumption that 'our understanding is just the same [as the infinite understanding], only in a limited way' (Maimon, *Gesammelte Werke*, vol. 2, 65; Maimon, *Essay on Transcendental Philosophy*, 38).[6]

Furthermore, if the limited human understanding is *just the same* as the infinite understanding then it follows that the former's transcendental apperception can, *contra* Kant, justify the claim that it is an apperceptive monad.

The continental rationalist aspect of his thinking is shaped by both Spinoza *and* Leibniz, and the influence of the latter is at least as great – if not greater – as the influence of the former. Indeed, when describing how the world exists for an understanding of which ours is only a part, Maimon writes that 'this is ... the point at which Leibnizians [and] Spinozans ... can be united' (Maimon, *Gesammelte Werke*, vol. 2, 208; Maimon, *Essay on Transcendental Philosophy*, 110).
[5]See also AA vol. 5, 401–403. For a detailed account of Kant's conception of intellectual intuition see Förster, The Twenty-Five Years of Philosophy, 141–152.
[6]In his 'Concerning the Progress of Philosophy' Maimon suggests that in mathematical reasoning we can attain some insight into the operation of such a divine understanding – and explicitly equates his views on this matter with the true 'spirit' of Leibniz. He writes:

> God ... does not think *discursively* as we do, rather his thoughts are at the same time presentations [*Darstellungen*]. If one objects that we have no concept of such a manner of thinking then I answer, we do actually have such a concept since we ourselves partially possess it. All *concepts of mathematics* are *thought* by it and at the same time presented as *real objects* through construction *a priori*. We are therefore in this respect similar to *God* [...]. God ... thinks *all real objects* ... (although certainly in a more complete manner) as we think the *objects of mathematics*, i.e. he brings them about at the same time *by his thinking*.
>
> (Maimon, *Gesammelte Werke*, vol. 4, 42)

For Kant, although I must represent myself as one and the same in my synthesis of the manifold of intuition, this mere representation '*I think*' accompanying and logically preceding all intuitions in no way justifies the claim that I am a substance that is simple, numerically identical and enduring throughout time. On Maimon's hypothesis that representation and thing are one and the same, however, Kant's objections to rational psychology fall aside, thus allowing Maimon to anticipate Fichte's subsequent claim (four years later) that, effectively, we do possess an 'intellectual intuition' of the 'I'.[7] Maimon thus writes:

> I maintain that the *I* is an intuition, indeed even an *a priori* intuition (because it is the condition of all thought in general). So the category of substance can be applied to it and the question *quid juris?* does not arise. But if we inquire further: how do I recognize the fact that my I endures in time? Then I reply: because it accompanies all of my representations in a time series. How do I recognize the fact that it is *simple*? Because I cannot perceive any manifoldness in it. How that it is numerically identical? Because I have cognition of it at different times as one and the same as itself. Kant no doubt objects that perhaps ... this is correct only of our representation of the I, but not with respect to this real thing that grounds it. But I have already explained that I hold the representation or concept of the thing to be one and the same as the thing, and that they can only be distinguished through the completeness of the latter with respect to the former. So, where there is no manifoldness, as in this case, the thing itself is one and the same as its representation, and what is valid of the latter must be valid of the former.
> (Maimon, *Gesammelte Werke*, vol. 2, 209–210; Maimon, *Essay on Transcendental Philosophy*, 111)

The understanding of such an apperceptive monad is limited, however, because in its consciousness of all things other than itself, its representations only partially correspond to – and are hence 'incomplete' in comparison with – things themselves. This occurs because consciousness is always necessarily related to a manifold which it must synthesize. Maimon thus writes that 'the condition of our thought (consciousness) in general is unity in the manifold' (Maimon, *Gesammelte Werke*, vol. 2, 16; Maimon, *Essay on Transcendental Philosophy*, 13) since 'without synthesis no consciousness

---

[7]Fichte refers to the immediate intuition that he believes constitutes I-consciousness (i.e. the consciousness that *I am*) as intellectual intuition within his review of *Aenesidemus* of 1794 and *Versuch einer neuen Darstellung der Wissenschaftslehre* of 1797/98. He employs this term to capture how, in I-consciousness, there is no distinction between the *intuited* and *intuitant*. It seems that a case could be made for claiming that Fichte follows Maimon (for whom he expressed admiration) in claiming that, in regard to I-consciousness, there is – in one sense – no distinction between representation and thing. We must, however, add the caveat that Fichte himself resisted the ascription of 'thinghood' or 'substantiality' to the ultimate nature of subjectivity. For further information about the relationship between Maimon and Fichte, see Freudenthal, *Salomon Maimon*, 233–248.

is possible' (Maimon, *Gesammelte Werke*, vol. 2, 349; Maimon, *Essay on Transcendental Philosophy*, 181). Any actual and determinate object of consciousness must therefore ultimately consist of different 'elements', which, *qua* 'elements', cannot themselves be objects of consciousness and are thus appropriately referred to as 'ideas'. To distinguish them from Kant's 'ideas of reason', Maimon calls them 'ideas of understanding' and says that they are distinguished from the former on account of the fact that 'we must presuppose their existence in us for any determined consciousness' (Maimon, *Gesammelte Werke*, vol. 2, 350; Maimon, *Essay on Transcendental Philosophy*, 181). A straight line of determinate length can thus be an actual object of consciousness for the human understanding, but the human understanding can never attain consciousness of the infinitesimally small elements within which such a straight line must ultimately be resolved (through differentiation). For us such infinitesimals are only 'ideas' or 'limit concepts'. As such our understanding remains 'incomplete'. However, the infinite understanding's concept of a straight line and the straight line itself are one and the same. Indeed, the infinite understanding's concept of a straight line is, for Maimon, that very infinitesimally small element from which our imagination constructs through continual repetition (integration) any determinate straight line. And it is precisely this metaphysical picture which Maimon uses to answer the question *quid juris?* in respect of pure mathematics. Because 'the infinite thinking being thinks all possible concepts all at once with the greatest completeness without any admixture of sensibility', for this understanding the proposition that, for example, 'the straight line between two points is always the shortest' is analytic (Maimon, *Gesammelte Werke*, vol. 2, 183; Maimon, *Essay on Transcendental Philosophy*, 98). The limited human understanding can, however, merely be conscious of determinate straight lines and is 'not conscious of the concepts contained within sensibility' (Maimon, *Gesammelte Werke*, vol. 2, 182; Maimon, *Essay on Transcendental Philosophy*, 98). As such, although we can continually exhibit within intuition the apodictic truth that 'the straight line between two points is always the shortest', for us this truth remains synthetic. Maimon thus writes that 'the faculty of intuition ... *conforms to rules*, but does not *comprehend rules*' (Maimon, *Gesammelte Werke*, vol. 2, 34–35; Maimon, *Essay on Transcendental Philosophy*, 23). Against Kant, therefore, he argues that spatio-temporal intuition cannot ground the apodictic certainty of mathematical propositions, and that our need to appeal to such intuition simply demonstrates our imperfect comprehension of their true ground. He thus writes that 'synthetic propositions ... derive their existence from the incompleteness of our concepts' (Maimon, *Gesammelte Werke*, vol. 2, 65; Maimon, *Essay on Transcendental Philosophy*, 38) and that 'for us sensibility is incomplete understanding' (Maimon, *Gesammelte Werke*, vol. 2, 183; Maimon, *Essay on Transcendental Philosophy*, 98). The fate of human consciousness is thus merely to be able to partially grasp, by means of intuition, the concepts thought by an infinite understanding. We can thus see that to resolve the question *quid juris?* Maimon believes it necessary to depart from the Kantian system and return to Leibniz. He thus writes:

> How can the understanding subject something (the given object) to its power (to its rules) that is not in its power? In the Kantian system, namely, where sensibility and understanding are two totally different sources of our cognition, the question is insoluble as I have shown; on the other hand in the Leibnizian-Wolffian system, both flow from one and the same cognitive source (the difference lies only in the degree of completeness of the cognition) and so the question is easily resolved.
> (Maimon, *Gesammelte Werke*, vol. 2, 63–64; Maimon, *Essay on Transcendental Philosophy*, 38)

Maimon also believes that the hypothesis he adopts to make comprehensible pure mathematical cognition, can also be employed within transcendental philosophy to explain how we can legitimately connect *a priori* concepts with empirical intuitions to constitute, even if not 'experience' in Kant's sense, our consciousness of empirical objects and the alterations they undergo.

Firstly, Maimon argues that, since the infinite understanding thinks all possible concepts, and, in doing so, all possible relations between them, it thinks of concepts (a) in terms of identity and difference, as (b) in a relation of dependency (in accordance with which one is thought as *determinable* and another as a *determination*) and as (c) within relations of reciprocal determination. The (b) synthesis of *determinable* and *determination* describes the relation between subject and predicate, in which we find one term (i.e. the *determinable*) that can be thought 'either in itself or in another synthesis' (Maimon, *Gesammelte Werke*, vol. 2, 84; Maimon, *Essay on Transcendental Philosophy*, 49) and another term (i.e. the *determination*) that can only be thought in reference to the former (i.e. the *determinable*). In the (c) synthesis of reciprocal determination, on the other hand, we find two terms which cannot be thought without reference to one another, where 'each is at the same time subject and predicate in relation to the other' (Maimon, *Gesammelte Werke*, vol. 2, 85; Maimon, *Essay on Transcendental Philosophy*, 50). Concerning (c) the latter, Maimon writes:

> an infinite understanding thinks all possible things by this means. For an infinite understanding everything is in itself fully determined because it thinks all possible real relations between the ideas as their principles. For example, let us suppose that x is a function of y, y a function of z, etc. A necessary relation of x to z etc. arises out of these merely possible relations. Through this new function, x is more determined than before, and through being related to all possible relations, it is completely determined.
> (Maimon, *Gesammelte Werke*, vol. 2, 86–87n; Maimon, *Essay on Transcendental Philosophy*, 50n)

This clearly parallels Leibniz's claims that God thinks all monads within a pre-established harmony and that by means of this 'each simple substance has relations that express all the others' (Leibniz, *Philosophical Essays*, 220). Furthermore, Maimon's claim that, for our finite understanding, the (c) synthesis of reciprocal determination 'is a mere form that (viewed in itself and apart from its application to a determined object of intuition) does not determine any object' (Maimon, *Gesammelte Werke*, vol. 2, 86n; Maimon, *Essay on Transcendental Philosophy*, 50n) clearly parallels Leibniz's claim that 'in simple substances the influence of one monad over another can only be ideal, and can only produce its effect through God's intervention' (Leibniz, *Philosophical Essays*, 219).

Secondly, Maimon argues that these very same concepts (which exist within the three aforementioned relations) serve as infinitesimally small elements or differentials from which the imagination of a limited understanding constructs empirical intuitions (through a process akin to mathematical integration). It is thus the repetition and synthesis of these elements, in accordance with spatio-temporal forms, that generates our empirical intuitions. But in order for us to understand these empirical intuitions as perceptions of real determinate objects subject to alteration, it is necessary for us to apply the categories of substance and accident and cause and effect. For Maimon, the understanding's application of categories is perfectly legitimate because it applies them neither to the empirical intuitions as such nor to their spatio-temporal structure, but rather to the elements from which these intuitions are constituted. And since these elements are concepts for an infinite understanding and our categories are reflections of the very same relations in which the infinite understanding thinks all concepts – the (b) synthesis of *determinable* and *determination* corresponding to our concepts of substance and accident and the (c) synthesis of reciprocal determination corresponding to our concept of cause and effect – Kant's problem concerning the heterogeneity of the *subsumed* and the *subsuming* falls away, so that thought and intuition can be here said to be quite legitimately bound together. Maimon thus writes:

> The metaphysically infinitely small is real because quality can certainly be considered in itself abstracted from all quantity. This way of considering it is also useful for resolving the question *quid juris?* because the pure concepts of the understanding or categories are never directly related to intuitions, but only to their elements, and these are ideas of reason [i.e. ideas of understanding] concerning the way these intuitions arise; it is through the mediation of these ideas that the categories are related to the intuitions themselves. Just as in higher mathematics we produce the relations of different magnitude themselves from their differentials, so the understanding (admittedly in an obscure way) produces the real relations of qualities themselves from the real relations of their differentials. So, if we judge that fire melts wax, then this judgement does not relate to fire and wax as objects of intuition, but to

their elements, which the understanding thinks in the relation of cause and effect to one another.
(Maimon, *Gesammelte Werke*, vol. 2, 133; Maimon, *Essay on Transcendental Philosophy*, 74)

Maimon's Leibnizian claim that sensibility is incomplete understanding inevitably leads him to also promulgate a Leibnizian conception of time and space, in which these 'forms of intuition' are, *contra* Kant, conceived as 'concepts of the understanding of the connections and relations of things in general' (Maimon, *Gesammelte Werke*, vol. 2, 64; Maimon, *Essay on Transcendental Philosophy*, 38). He thus writes that 'I speak ... as a Leibnizian who treats time and space as universal undetermined concepts of reflection that must have an objective ground' (Maimon, *Gesammelte Werke*, vol. 2, 133; Maimon, *Essay on Transcendental Philosophy*, 74). This objective ground is the manifold of concepts or things existing within an infinite understanding. Since human thought or consciousness is a 'synthesis', it only emerges insofar as the aforementioned elements are conceived by the understanding as different, and are thus subsumed under the understanding's pure concept of difference. In order that this difference can be perceived, however, these different elements must be 'ordered' by the imagination. That is to say, repetition or integration must occur so that these different elements become, for us, different sensible objects. This ordering process, however, universally occurs within a spatio-temporal form. Much follows from this: Firstly, it follows that, *contra* Kant, we should not conceive of space and time as pure forms of intuition which logically *precede* the sensible data organized within them, but rather conceive of the space and time which we intuit as arising *as a result* of the imagination's organization of that sensible data; and secondly, it follows that any perceptual field must contain a diversity of perceptual data in order for us to be conscious of it at all. Both points are illustrated by the following passage:

> Kant ... only proves that space is a universal concept, not that it is an *a priori* one (according to my definition). By contrast, I claim that as intuition space is a schema or image of the difference between given objects, that is, a subjective way of representing this objective difference, a difference that is a universal form or necessary condition of thinking things in general; without this objective difference it would be an empty space, i.e., a transcendent representation without any reality (as when I imagine an homogenous object in space, without relating it to something heterogeneous). So, considered in itself, space is indeed a universal concept but not an *a priori* one; only when considered in relation to what it represents (difference) is it an *a priori* concept, namely because difference pertains to all things; or all things must be – or must be thought as – different from one another: for it is just this that makes them all things.
> (Maimon, *Gesammelte Werke*, vol. 2, 179–180; Maimon, *Essay on Transcendental Philosophy*, 96–97)

Space and time are therefore originally concepts of the difference between things, but Maimon claims that one can concede this point without necessarily falling foul of Kant's objections to this Leibnizian conception, by stressing that the intuited space and time generated by the imagination (as the schema of the concept of difference) are just as universal as they were for Kant.[8] Of course, *qua* intuitions, space and time remain, for Maimon, just as subjective as they were for Kant, although they nevertheless possess a real or objective ground which would remain even if our capacity for intuition were annihilated. As such, Maimon describes his position as one that lies between the extremes of transcendental idealism and transcendental realism (Maimon, *Gesammelte Werke*, vol. 2, 204–206; Maimon, *Essay on Transcendental Philosophy*, 108–109): He is not a transcendental realist insofar as he is certain that no *determined* objects could exist independently of his faculty of intuition, but he is not a transcendental idealist insofar as he does not deny that something *determinable* could exist independently of his faculty of intuition. For, divorced from this faculty, the ideas thought by the understanding would still remain. And it is precisely such an intermediary position which he claims to find in Leibniz. He thus writes that 'I could ... easily show that [my] system agrees very precisely with the Leibnizian one (if this system is correctly understood)' (Maimon, *Gesammelte Werke*, vol. 2, 206; Maimon, *Essay on Transcendental Philosophy*, 110).

We thus see how Maimon reconciles Leibniz's monadology with Kantian criticism. Leibniz's God is clearly conceived by Maimon as a being possessing 'intellectual intuition' in the Kantian sense. Therefore, for Maimon, any distinction between God's *concepts* and God's *actualized possibilities* is merely a distinction of reason. Accordingly, Maimon's monads are not things in themselves, but concepts contained within understanding, which although grasped completely by God's understanding, are nonetheless grasped incompletely by us. Hence, within our limited understanding, these monads constitute the infinitesimally small elements from which our perceptions emerge. And therefore, as Leibniz stressed, the empirical content of our perceptions can in no way be accounted for in terms of a causal affection on the part of things independent of us. Likewise, according to Maimon, the same denial of the affective origin of the empirical content of perceptions is present within Kant:

> Kant very often uses the word 'given' in connection with the matter of intuition; by this he does not mean (and nor do I) something within us that has a cause outside us ... hence 'given' signifies only this: a representation that arises in us in an unknown way.
> (Maimon, *Gesammelte Werke*, vol. 2, 203; Maimon, *Essay on Transcendental Philosophy*, 108)

---

[8]In an insightful discussion of Maimon's conception of space and time, Peter Thielke notes that Maimon's 'position escapes the obvious objections Kant levels against the Leibnizian explanation of space and time, while retaining the general tenor of the rationalist party line' (Freudenthal, *Salomon Maimon*, 90).

Maimon adds that since 'space itself is only a form within us' (Maimon, *Gesammelte Werke*, vol. 2, 203; Maimon, *Essay on Transcendental Philosophy*, 108) it would be incoherent to talk of a cause outside us in a spatial sense – and about this Kant would certainly agree (see, e.g. A373). He goes on to argue, however, that since time is likewise within us, it would be incoherent to employ the category of causality to even *think* the source of the material content of our intuitions. Accordingly, for Maimon, there exist only concepts and perceptions within an understanding, and it would be incoherent to even talk of things in themselves supplying the given material content of perception:

> For Kant, the given representation cannot signify what has a cause outside the faculty of representation because it is unthinkable that we can have cognition of the thing in itself (noumenon) as a cause outside the faculty of representation, since in this case the schema of time is lacking.
> (Maimon, *Gesammelte Werke*, vol. 2, 415/419; Maimon, *Essay on Transcendental Philosophy*, 213)

For Maimon, therefore, 'differentials of objects are the so-called *noumena*' (Maimon, *Gesammelte Werke*, vol. 2, 32; Maimon, *Essay on Transcendental Philosophy*, 21).

3. SCHELLINGIAN APPROPRIATIONS

In 1797 Schelling can also be seen as reconciling Kantian criticism with Leibnizian monadalogy. Schelling is well-known for his constant changes in philosophical position and also for the affinity of his thought with that of Spinoza. When considering his career as a whole, one would inevitably have to conclude that he was far more of a Spinozist than a Leibnizian. Nevertheless, around 1797, Leibniz was at least as – if not more – important for him.

While Schelling's earliest works do not betray an especially high regard for Leibniz (see Schelling, *Historisch-kritische Ausgabe*, vol. I, 2, 110 & 140–145), in a letter to Jacob Hermann Obereit of 12 March 1796, Schelling is decidedly critical of him:

> I believe that Leibniz actually marks the beginning of philosophy's middle-ages (although the Scholastics had already paved the way for this), because in philosophy one then began to make the Absolute into a merely abstract being, and to consider God, not as the being *of all* beings, but rather (in the popular manner) as being *outside of* all beings.
> (Schelling, *Historisch-kritische Ausgabe*, vol. III, 1, 46)

This negative assessment was a response to Obereit's far more enthusiastic reception of Leibniz, who he saw as a precursor of Schelling's own position

(Schelling, *Historisch-kritische Ausgabe*, vol. III, 1, 36–37). It is possible that Obereit nevertheless provoked Schelling soon afterwards to enter into a more thorough study of Leibniz and to reconsider his relation towards him.[9] In any case, Schelling's 'General Overview' announces that 'the time has come to understand Leibniz. For surely he must not be understood as he has been understood *thus far*' (Schelling, *Historisch-kritische Ausgabe*, vol. I, 4, 170; Pfau, *Idealism and the Endgame of Theory*, 132). Leibniz's importance for Schelling during this period is also reflected in a letter to Niethammer of 4 August 1797, in which Schelling discusses a proposed work entitled *Philosophische Parallelen*, which he says would 'begin with an interpretation of the Leibnizian philosophy' (Schelling, *Historisch-kritische Ausgabe*, vol. III, 1, 134), as well as his claim within his *Ideas for a Philosophy of Nature* that 'the time has come when [Leibniz's] philosophy can be re-established' (Schelling, *Historisch-kritische Ausgabe*, vol. I, 5, 77; Schelling, *Ideas for a Philosophy of Nature*, 16). The question concerning Maimon's influence on Schelling during this period is difficult to determine. Within *Vom Ich als Princip* of 1795, Schelling praises Maimon for his critical destruction of Reinhold's *Grundsatzphilosophie* (Schelling, *Historisch-kritische Ausgabe*, vol. I, 2, 137; see also: Schelling, *Historisch-kritische Ausgabe*, vol. I, 1, 267) and his penetrating analysis of the nature of infinite judgements (Schelling, *Historisch-kritische Ausgabe*, vol. I, 2, 151), thus proving his engagement with Maimon's *Versuch einer neuen Logik oder Theorie der Denkens* of 1794. Although it is most probable that he knew of it, there is no explicit evidence to suggest that Schelling similarly engaged with the *Essay*. Nonetheless, we can still find remarkable systematic parallels between Maimon's position in 1790 and Schelling's position in 1797.

In the 'General Overview', Schelling explicitly states that its interpretation of Kant is not intended to present what Kant himself had actually 'intended with his philosophy but merely [to provide] what ... he *had* to have intended if his philosophy was to prove internally cohesive' (Schelling, *Historisch-kritische Ausgabe*, vol. I, 4, 107; Pfau, *Idealism and the Endgame of Theory*, 84). Like Maimon, Schelling argues that if Kant is understood as a cognitive dualist – asserting 'an utter separation of the understanding and sensibility' (Schelling, *Historisch-kritische Ausgabe*, vol. I, 4, 77; Pfau, *Idealism and the Endgame of Theory*, 73)[10] – then his philosophy

---

[9]That Schelling entered into a study of Leibniz within this period has been suggested by Manfred Duner. See Schelling, *Historisch-kritische Ausgabe*, vol. I, 5, 16–17.

[10]Following this passage, the editors of the *Historisch-kritische Ausgabe* invite comparison with §697 of Ernst Platner's *Philosophische Aphorismen nebst einigen Anleitungen zur philosophischen Geschichte* of 1793. In this section, which is headed '*throughgoing separation of the sensibility from the understanding and the intuition from the concept*' (as qtd. in Schelling, *Historisch-kritische Ausgabe*, vol. I, 4, 315), Platner questions the grounds for Kant's cognitive dualism, and within this context refers to Maimon's *Essay*, writing: 'Maimon, otherwise a friend of the Kantian system, finds this separation unnatural; *Transcendental Philosophy*, pp. 63, 183' (as qtd. in Schelling, *Historisch-kritische Ausgabe*, vol. I, 4, 316).

cannot be 'internally cohesive'. As Schelling notes, Kant's cognitive dualism amounts to the claim that 'the *form* of our cognitions originates within *ourselves*, whereas its *matter* is given to us *from the outside*' (Schelling, *Historisch-kritische Ausgabe*, vol. I, 4, 82; Pfau, *Idealism and the Endgame of Theory*, 76). Echoing Maimon's criticisms of Kant's answer to the question *quid juris?*, Schelling proceeds to argue that, on this assumption, it remains incomprehensible as to how the understanding's categories can be legitimately applied to the data passively received by sensibility. He writes:

> For [Kant's disciples] the world and all reality prove primordially alien to our spirit [*Gesite*], and the world bears no affinity to the spirit other than that of an *accidental* [*zufällige*] affect. Nevertheless *such* a world, although for them it is merely *accidental* and thus might just as well be different, they claim to govern with laws that – they neither know how nor whence – have been implanted in their understanding. As the supreme legislators of nature [and] with the full consciousness that the world is composed of things in themselves, they impose these concepts and laws of understanding onto these things in themselves... and this world of eternal and determinate nature obeys their speculative decree... A more ridiculous or preposterous system has hardly ever been thought out.
> (Schelling, *Historisch-kritische Ausgabe*, vol. I, 4, 78–79; Pfau, *Idealism and the Endgame of Theory*, 74)

Schelling then proceeds to argue that, because such a system would render contingent the affinity between the world and the understanding's laws, it could not prove that we possess cognition of a law-governed nature as opposed to a merely subjective sequence of perceptions, which, due to laws of association, gives us the illusion of insight into a law-governed reality. Thus, also like Maimon, Schelling raises the question *quid facti?* and, in this regard, evokes Hume's scepticism:

> Our understanding is supposed to have imposed [these laws] onto nature as something completely heterogeneous. *Hume*, the sceptic, first had claimed what is now being attributed to Kant. Yet Hume readily admitted that all natural sciences amount to deception, [and] that all laws of nature constitute but a routine [*Gewohnheiten*] of our imagination. This was a consistent philosophy. And Kant is supposed to have done no more than repeat Hume so as to now render him, who had been consistent, inconsistent?
> (Schelling, *Historisch-kritische Ausgabe*, vol. I, 4, 79; Pfau, *Idealism and the Endgame of Theory*, 78)

Schelling is thus convinced that, if transcendental philosophy is intended to account for the legitimacy of our claims to possess *a priori* cognition of nature, it cannot proceed from the standpoint of cognitive dualism.

It is in the process of reconfiguring transcendental philosophy to exclude such pernicious dualisms that Schelling has recourse to the resources of Leibnizian philosophy. Whereas Maimon, in order to explain the possibility of *a priori* cognition, had recourse to an 'infinite understanding', Schelling has recourse to – what he describes as – an 'infinite spirit' (Schelling, *Historisch-kritische Ausgabe*, vol. I, 4, 79; Pfau, *Idealism and the Endgame of Theory*, 78) or 'absolute *subject*' (Schelling, *Historisch-kritische Ausgabe*, vol. I, 4, 85; Pfau, *Idealism and the Endgame of Theory*, 78). This infinite spirit acts upon itself to become a finite object for itself, or, in other words, intuit itself (Schelling, *Historisch-kritische Ausgabe*, vol. I, 4, 141; Pfau, *Idealism and the Endgame of Theory*, 112). *Qua intuiting*, spirit is infinite and expansive; *qua intuited*, spirit is finite and contractive. An 'absolute *simultaneity* of the infinite and the finite' writes Schelling 'contains the *essence* of an *individual* nature (of selfhood [*Ichheit*])' (Schelling, *Historisch-kritische Ausgabe*, vol. I, 4, 86–87; Pfau, *Idealism and the Endgame of Theory*, 79. See also Schelling, *Historisch-kritische Ausgabe*, vol. I, 5, 91; Schelling, *Ideas for a Philosophy of Nature*, 16). And this individual 'I' (the product of spirit's self-intuition) is explicitly conceived as a windowless monad. Schelling thus tells us that 'the human spirit is of an *organic nature*, nothing will enter it *mechanically from the outside*' (Schelling, *Historisch-kritische Ausgabe*, vol. I, 4, 113; Pfau, *Idealism and the Endgame of Theory*, 92) and that 'primordially, we possess neither an intuition of things *outside ourselves* nor, as some have taught, in *God*, but ... we intuit them exclusively *within ourselves*' (Schelling, *Historisch-kritische Ausgabe*, vol. I, 4, 118; Pfau, *Idealism and the Endgame of Theory*, 95). Schelling thus claims that infinite spirit produces finite spirits which intuit a world within themselves, in an analogous manner to how Leibniz claimed that God produces monads with their own perceptions of a world. Indeed, Schelling explicitly equates his view that 'the world itself consists only of this expansion and contraction of the spirit' with Leibniz's 'image of a *perpetual creation*' (Schelling, *Historisch-kritische Ausgabe*, vol. I, 4, 123; Pfau, *Idealism and the Endgame of Theory*). In re-appropriating this Leibnizian position, however, it is clear that Schelling believes that, if we are to focus upon the 'spirit' as opposed to the 'letter' of his philosophy, we must not understand the Leibnizian God as a 'being *outside of* all beings', but rather, like Maimon, see the apperceptive monad as a limitation of the infinite spirit – and, conversely, sees the divine as immanently contained within ourselves. That Schelling came to read Leibniz in such a manner is suggested by the following passage, which implies that Leibniz – just as much as Spinoza and Maimon – conceived of the human soul as a limitation of an immanent God:

> [Spinoza's] system was the first bold outline of a creative imagination, which went over from the infinite in idea to the finite in intuition. *Leibniz* came, and went the opposite way. [...] The first thought from which he set out was: 'that

the representations of external things would have arisen in the soul by virtue of her own laws *as in a particular world*, even though nothing were present but God (the infinite) and the soul (the intuition of the infinite)'.
(Schelling, *Historisch-kritische Ausgabe*, vol. I, 5, 76–77; Schelling, *Ideas for a Philosophy of Nature*, 15–16)

Schelling also argues that it is only insofar as God is conceived in such a way – as a being who creates the world *through* our*selves* – that we can explain our cognition of this world:

> even if we understand the origin of a world *external to ourselves*, we still do not understand how the representations of this world could have entered *into* our consciousness. In the last effort, then, it had to be explained not how external things could have originated independently of ourselves – (for of these we cannot have any knowledge, since they themselves are the ultimate substratum for any explanation of external phenomena) – but how a representation of these [things] could have originated *within* us.
> (Schelling, *Historisch-kritische Ausgabe*, vol. I, 4, 83; Pfau, *Idealism and the Endgame of Theory*, 77)

If the world consists of things in themselves created by a transcendent God, accounting for the possibility of *a priori* cognition of the world would indeed embroil the Kantian cognitive dualist within the aforementioned problems Schelling discusses. Adopting the Leibnizian view, however, the problem falls away. Accordingly, for Schelling, 'nature is nothing different from [its] laws' (Schelling, *Historisch-kritische Ausgabe*, vol. I, 4, 79; Pfau, *Idealism and the Endgame of Theory*, 75). Similarly, by adopting this Leibnizian position, Schelling, like Maimon, also rejects the Kantian view that 'we ... simply *bring along* [the] forms [of space and time] as something finished and ready made for the purpose of intuition' (Schelling, *Historisch-kritische Ausgabe*, vol. I, 4, 73; Pfau, *Idealism and the Endgame of Theory*, 71). Space and time are instead declared to be 'modes of activity of our subject' (Schelling, *Historisch-kritische Ausgabe*, vol. I, 4, 74; Pfau, *Idealism and the Endgame of Theory*, 72) – the former corresponding to the infinite and expansive, the latter corresponding to the finite and contractive – in accordance with which the transcendental imagination generates intuitions of finite objects. Thus, anything that is an object exists only for a subject, for objects are generated by the imagination's synthesis of that which is created *ex nihilo* through the infinite spirit's original act of self-intuition. Accordingly, also as in Maimon, the material content of intuition is not to be explained by the affection of things in themselves:

> If [Kant's doctrine of the transcendental synthesis of imagination] had been understood, the chimera that has tormented our philosophers for so long – viz. the things in themselves ... would have disappeared like mists of the

night dispelled by the light of the sun. It would have been recognized that nothing can be real unless there is a spirit to know it. For Leibniz 'things in themselves' were something quite different. Leibniz did not know of any other being other than one that *knows itself* or *is known* by a spirit [*das sich selbst erkennt oder von einem Geiste erkannt wird*]. The latter he considered strictly as appearance. Yet he did not turn into a dead, selfless object what exceeded mere appearance. For that reason he invested his monads with the capacity for representation and turned them into mirrors of the universe, into knowing [*erkennenden*], representing, and precisely *to that extent* not 'knowable [*erkennbaren*]' and 'representable' beings.

(Schelling, *Historisch-kritische Ausgabe*, vol. I, 4, 75–76; Pfau, *Idealism and the Endgame of Theory*, 72)

Therefore, for Schelling, just as little as for Maimon, Leibniz's monadology is not to be understood as a metaphysics resolving the world into an infinity of simple things in themselves. Hence, according to Schelling, a focus upon the 'spirit' (as opposed to the 'letter') of the work of Leibniz and Kant reveals that both systems can be reconciled.

Kant may well have understood Leibniz's monads as things in themselves thought by means of transcendental ideas, but, as we have seen, some of his successors provided other interpretations. Maimon and Schelling introduce monadologies as necessary hypotheses for explaining our capacity for *a priori* cognition of the world, with the result that their monads could be said to likewise be transcendental conditions of the possibility of our knowledge and experience. For both Maimon and Schelling, therefore, Leibnizian monadology can not only survive the onslaught of Kantian criticism, but even provides the latter with a vitally important source of support.

BIBLIOGRAPHY

Förster, Eckart. *The Twenty-Five Years of Philosophy*. Cambridge, MA: Harvard University Press, 2012.
Freudenthal, Gideon, ed. *Salomon Maimon: Rationalist Dogmatist, Empirical Skeptic*. Dordrecht: Kluwer, 2003.
Kant, Immanuel. *Gesammelte Schriften*. Edited by Königlich Preußische Akademie der Wissenschaften, Akademie der Wissenschaften zu Berlin, and Akademie der Wissenschaften zu Göttingen. Berlin: de Gruyter, 1900ff.
Kant, Immanuel. *The Cambridge Edition of the Works of Immanuel Kant*. Edited by Paul Guyer and Allen W. Wood. Cambridge: Cambridge University Press, 1992ff.

Leibniz, G. W. *Philosophical Essays*. Translated by Roger Ariew and Daniel Garber. Indianapolis: Hackett, 1989.
Maimon, Salomon. *Gesammelte Werke*. Edited by Valerio Verra. Hildesheim. Zürich: Georg Olms Verlag, 2003.
Maimon, Salomon. *Essay on Transcendental Philosophy*. Translated by Nick Midgley, Henry Somers-Hall, Alistair Welchman, and Merten Reglitz. London: Continuum, 2010.
Melamed, Yitzhak Y. 'Salomon Maimon and the Rise of Spinozism in German Idealism'. *Journal of the History of Philosophy* 42, no. 1 (2004): 67–96. doi:10.1353/hph.2004.0010.
Pfau, Thomas, ed. *Idealism and the Endgame of Theory: Three Essays by F. W. J. Schelling*. Albany: State University of New York Press, 1994.
Schelling, Friedrich Wilhelm Joseph. *Historisch-kritische Ausgabe*. Edited by Jörg Jantzen, Thomas Buchheim, Jochem Hennigfeld, Wilhelm G. Jacobs, and Siegbert Peetz. Stuttgart: Frommann-Holzboog, 1976ff.
Schelling, Friedrich Wilhelm Joseph. *Ideas for a Philosophy of Nature*. Translated by Errol E. Harris and Peter Heath. Cambridge: Cambridge University Press, 1988.

ARTICLE

HERBART'S MONADOLOGY

Frederick Beiser

This article is an introduction to Herbart's monadology. It discusses the fundamental concepts of his monadology and its similarity to Leibniz's monadology. A final section discusses the vexed question of Herbart's realism. It is argued that Herbart is more a transcendental idealist than a realist.

1. HISTORICAL SIGNIFICANCE

In the history of post-Kantian monadology, a prominent place should be given to Johann Friedrich Herbart (1776–1841). Along with Fichte, Schelling and Hegel, who were his contemporaries, Herbart was one of the great metaphysicians of the early nineteenth century. His metaphysics was a monadology in the most basic sense: it postulated the existence of a multitude of simple animate entities to explain the world. Although Herbart did not use the term 'monadology' to describe his metaphysics – the term had too many troubling associations by the early nineteenth century – and although he had some major disagreements with Leibniz, he still expressly saw his metaphysics as a rehabilitation of Leibnizian doctrine. His *Allgemeine Metaphysik* begins with the telling lines: 'If there were any point in building a splendid door for the entrance to this work, we would find the materials for it from no one better than Leibniz' (§1; VII, 21).[1] As we shall soon see, however, Leibniz provided more than the building materials for the entrance to Herbart's metaphysics; he also supplied its foundation stones.

A contemporary reader might ask why he or she should bother with Herbart's monadology. It seems to be a hopeless attempt to revive antiquated doctrine, one thoroughly discredited by Kant's critique of metaphysics.

---

[1] All references to Herbart will be to *Sämtliche Werke*. References to this edition will be to volume (Roman numeral) and page number (Arabic numeral). References to the *Allgemeine Metaphysik* will be first to the paragraph number (§) and then the volume and page number. 'Anm.' designates 'Anmerkung', a remark appended to a paragraph. This work consists in two volumes, volumes 7 and 8 of *Sämtliche Werke*. Volume 7 consists in paragraphs §§1–160 and volume 8 in paragraphs §§161–444. All translations from the German are my own.

But if we adopt this dismissive attitude, we only beg the question against Herbart. For he was well aware of the challenge that Kant had posed to metaphysics, and he was very concerned to meet it. Herbart even described himself as 'ein Kantianer von 1828',[2] and he insisted that 'any new metaphysics that comes forward as a science alone' has to meet Kantian standards. Hence he understood metaphysics as 'die Wissenschaft der Begreiflichkeit der Erfahrung', as 'ars experientiam recte intelligendi'. The task of metaphysics, in other words, was to determine the conditions of the possibility of experience. The first volume of Herbart's *Allgemeine Metaphysik* is a long discourse about what went right and what went wrong with not only pre-Kantian but also post-Kantian metaphysics; it also contains a diagnosis of the failure of the critical revolution itself. Herbart's central argument is that metaphysics can survive in the post-Kantian age only if it reinvents itself as a *critical* monadology. If we are to be fair to Herbart – and if we are to give a fair hearing to monadology itself – we do well to ponder that argument.

Apart from his monadology, any student of metaphysics has strong historical and philosophical reasons to take an interest in Herbart. For, though it has been nearly forgotten, Herbart's metaphysics was the contemporary of, and competitor to, that of the German idealists (Fichte, Schelling and Hegel) and that of the early romantics (Friedrich Schlegel, Novalis, Hölderlin and Schleiermacher). While the idealists and romantics found their inspiration in Plato and Spinoza, Herbart found his in Leibniz and Kant. Their metaphysics was monistic, teleological and idealistic (in the Platonic sense); Herbart's metaphysics was pluralistic, mechanistic and nominalist. A stronger antithesis is hardly imaginable. So, if we are interested at all in the philosophical basis of idealism and romanticism, we would do well to consider Herbart's criticism of these traditions. We understand and appreciate any philosophical tradition only when we know its antithesis.

Herbart's important historical place in the early nineteenth century raises the question: How could he be so forgotten? There are countless studies of the German idealists, singly and collectively; and there is a growing amount of work on the philosophy of the early romantics. But it is difficult to find today a solid and serious monograph on Herbart.[3] Why? It is tempting

---

[2]See the 'Vorrede' to *Allgemeine Metaphysik*, VII, 13. See also Herbart's academic lecture *Oratio ad capessendam in academia georgia augusta professionem philosophiae ordinariam habita*, SW X, 53–64, where he states (63): 'Kantianum ipse me professus sum, atque etiam nunc profiteor ... '.

[3]A computer search through the WorldCat for all publications on Herbart from 1950 until 2014 yielded 1462 items, though almost all of them were new printings or editions of his works or studies of his theory of education. Only a handful, which for reasons of space we cannot mention here, were on his philosophy proper. For a bibliography of works on Herbart before this period, which were more numerous, see Schmitz, *Herbart Bibliographie*. The best treatment of Herbart's metaphysics is still that of Cassirer, *Das Erkenntnisproblem in der Philosophie*, III, 378–410.

to pin some of the blame for this on Herbart's difficult exposition, which places enormous demands on the reader's patience and understanding. Herbart said that he wrote for students already initiated into his philosophy, an audience which has long disappeared. But if troubling expositions alone were the reason for his neglect, the idealists and romantics would be just as forgotten. The reason for the ignorance of Herbart lies rather in an entrenched tradition of philosophical historiography. For generations, that tradition has stood under the shadow of Hegel's influential *Geschichte der Philosophie*,[4] which portrayed the history of philosophy after Kant as the story of the development of his own system. Hegel simply wrote much of his opposition out of that story, according his opponents either a half-page (in the case of the romantics) or simply leaving them out entirely (Fries, Herbart and Beneke). Hegel's history is a wonderful story about his own intellectual development; but we cannot take it seriously as a general history of philosophy. Yet the major nineteenth-century historians after Hegel – Karl Rosenkranz, Johann Erdmann and Kuno Fischer – were Hegelians and therefore happy to accept their master's story as gospel (see Rosenkranz, *Geschichte der kant'schen Philosophie*; Erdmann, *Die Entwicklung der deutschen Spekulation seit Kant*; Fischer, *Geschichte der neueren Philosophie*, vols 5–7). They were followed in the twentieth century by Richard Kroner and Frederick Copleston (Kroner, *Von Kant bis Hegel*; Copleston, *Fichte to Hegel*) who, though no Hegelians, were content to repeat the historiography before them. Recent histories of nineteenth-century philosophy have, by and large, simply followed the Hegelian tradition (see e.g. Solomon, *Continental Philosophy since 1750*; Sandkühler, *Handbuch Deutscher Idealismus*; Bréhier, *The Nineteenth Century*).

Our task in this article will be to make up for some of the historical deficit in our understanding of Herbart. We will attempt to outline the basis of his monadology and show how it arose from his dispute with earlier traditions. Finally, we will examine one of the most misleading and persistent myths about Herbart's philosophy: its alleged realism. It should be obvious that all we can do here will be introductory and rudimentary. A proper understanding and appreciation of Herbart is the task for a future generation.

2. FAILURES OF PAST METAPHYSICS

Herbart developed his metaphysics in self-conscious reaction against three opposing traditions: rationalism, idealism and romanticism. The rationalist tradition meant for him chiefly Leibniz, Descartes, Spinoza and Wolff, what we would today call pre-Kantian metaphysics. The idealist tradition,

---

[4]Hegel, *Vorlesungen über die Geschichte der Philosophie*, volumes 18–20 of *Werke*. The third section of volume 20, 314–462, 'Neueste deutsche Philosophie', treats what we would now call the post-Kantian systems.

which was in formation in Herbart's day, was for him chiefly the work of Kant, Fichte and the young Schelling; Hegel, whom Herbart saw as a Schellingian, received little of his attention.[5] The romantic tradition was typified for Herbart mainly by Schelling and Schleiermacher; he never mentions other romantics (viz. Friedrich Schlegel, Novalis, Hölderlin). To understand Herbart's metaphysics, we do well to consider his diagnosis of the failures of these traditions.

Herbart knew the idealist and romantic traditions well, for he had been for a long time in his youth an idealist; and he was, for a shorter time, even a romantic. His philosophical apprenticeship was under the idealism of Fichte. In his early Jena years (1794–97) he was regarded as Fichte's most talented disciple; but, gradually and painfully, he worked his way out of his teacher's legacy. The Fichtean ego, he realized by 1800, is only a fantasy (see his 'Kritik der Ichvorstellung', which was written in May 1800, in *Sämtliche Werke* I, 113–14). Already in 1806 he published the first version of his own metaphysics, *Hauptpuncte der Metaphysik*,[6] which lays down the basis for much of his later system. It took another two decades, however, for his system to mature. The chief exposition of that system is his *Allgemeine Metaphysik*, which first appeared in two volumes from 1828 to 1829 (reproduced in volumes 7 and 8 of *Sämtliche Werke*).

Herbart's final settling of accounts with the idealist and romantic traditions appears in the first volume of his *Allgemeine Metaphysik*, which is a critical survey of metaphysics from Leibniz to Schelling. Herbart makes three fundamental criticisms of the idealists, two of which also apply to the romantics.[7] *First, a faulty methodology*. The idealists, specifically Fichte and the young Schelling, inadvisedly followed Reinhold's method of beginning philosophy with a single self-evident first principle. It is absurd, Herbart argued, to limit the starting point of a philosophy down to a single proposition. We should begin enquiry from many different angles, and we should not limit ourselves down to one alone. No single proposition, in any case, is so fecund and replete with consequences that we can construct an entire system from it alone; we need at least several first principles to obtain more than trivial results. Even a set of first principles, however, is

---

[5] Herbart knew Hegel well enough. He made some passing remarks about his philosophy in the preface to the second volume of his *Allgemeine Metaphysik* (VIII, 6). He wrote a highly critical review of Hegel's *Philosophie des Rechts* for the *Leipziger Literatur-Zeitung* in 1822 (Nr. 45–47), *Sämtliche Werke* XII, 140–54. Herbart saw Hegel as part of the Schellingian school and saw that as the chief problem with his philosophy.

[6] A second edition appeared in 1808 in Göttingen with Justus Friedrich Danckwerts. Both editions are reproduced in *Sämtliche Werke*, II, 175–226.

[7] The romantics (Friedrich Schlegel, Novalis and Hölderlin) did not accept the idealist methodology of first principles; they were in their early years severe critics of Reinhold's *Elementarphilosophie*. Herbart's first criticism of the idealists' methodology would therefore not apply to them. The romantics did, however, accept teleology and intellectual intuition, so the second and third criticisms would hold also for them.

at best a means for organizing and systematizing our knowledge, Herbart insists, so that we should not confuse the order of exposition or knowledge, the *ratio cognoscendi*, with the order of being or things, the *ratio essendi*. Reinhold, Fichte and Schelling are guilty of just this fallacy, however, because they think that their first principle has not only epistemic but also ontological priority; they make the ego not only the first principle of knowledge but also the first principle of things (§98, VII, 174; §137, VII, 271). *Second, confusion of the normative and factual.* Because of their Reinholdian methodology, Fichte and Schelling sought for the unity of reason, a single principle to unite theoretical and practical reason. But, Herbart insisted, there is no such point of unity. The practical and the theoretical, the normative and the factual, are distinct forms of discourse which should not be confused with one another. No 'ought' ever follows from an 'is', no obligation ever flows from a fact. Because he sharply distinguished the normative from the factual, Herbart rejected teleology, which he regarded as a confusion of these forms of discourse. For similar reasons he questioned the organic concept of nature, which had been such an inspiration for Schelling and the young romantics. To assume that nature is an organism, he argued, is to presuppose that it has purposes or ends, that there is something that it should be. However, the best maxim of natural explanation is, as far as possible, to account for the mechanism of things. Teleological explanation is a mere stopgap until we have understood that mechanism (§160; VII, 324). *Third, intellectual intuition.* Seeing that the understanding or discursive intellect is limited to analysing concepts, and admitting that sensory intuition is restricted to sense experience, Fichte, Schelling and the romantics appealed to intellectual intuition as a source of knowledge. Allegedly, such an intuition gives us insight into an intelligible realm above or behind the realm of sense experience. Through it, we know either the absolute or the pure activity of the ego. But, for Herbart, the recourse to intellectual intuition is a desperate stratagem. It is difficult to persuade or convince anyone of the content of such an intuition if he or she does not have such a marvellous and magical faculty (§94; VII, 162). It is indeed almost impossible to communicate such intuitions; for to articulate them would be to describe them with concepts, which is to destroy their immediacy and simplicity. If we are to discuss and dispute, Herbart believed, there is no substitute for clear and distinct concepts. Philosophy is for him, as it was indeed for Kant, a discipline of conceptual thinking; hence he described it as 'Bearbeitung der Begriffe', i.e. the working through or analysis of concepts.[8]

---

[8]See Herbart's definition of philosophy in *Lehrbuch zur Einleitung in die Philosophie, Sämtliche Werke* IV, 38–9. The reference to work or 'Arbeit' in 'Bearbeitung' is deliberate. Famously, Kant had spoken of the need for 'Arbeit' in philosophy, which he understood as the labour involved in the analysis of concepts. He contrasted such work with a more lazy philosophy that would appeal to feelings and intuitions. See his 'Von einem neuerdings erhobenen vornehmen Ton in der Philosophie'. Herbart was siding with Kant against Schelling and the romantics.

Herbart's own metaphysics also derived from his diagnosis of the failures of pre-Kantian metaphysics. The chief fallacy of the old metaphysics, Herbart explains, was its conflation of essence with existence, i.e. its assumption that existence is only another property of a thing. The metaphysics of the Leibnizian–Wolffian school held that the actual or real is what is completely determinate (§5; VII, 23). The movement from the possible to the actual was then understood as one from the indeterminate to the determinate, as if to get to the existence of a thing one only had to add another property to it (§7; VII, 24–5). This conflation of essence and existence is also evident in Descartes and Spinoza, who, in one form or another, argue that God's existence follows of necessity from his essence. 'Quo plus realitatis, aut esse unaquœque res habet, eo plura attributa ipsi competunt' – those lines from Proposition IX, Part I, of Spinoza's *Ethica* were for Herbart the perfect epitome of the chief fallacy of rationalist metaphysics (Spinoza, *Opera-- Werke* II, 96. 'The more reality or being a thing has, the more attributes belong to it'). It was Kant's greatest contribution to the critique of metaphysics, Herbart claimed, to have shown the fallacy in this old way of thinking and to have sharply distinguished essence from existence (§32; VII, 55).

The net effect of this fallacy was that traditional metaphysics had become too removed from experience and existence. It assumed that its analysis of concepts was sufficient to understand reality, though there was no guarantee that these concepts had any reference at all.

A new metaphysics, Herbart was convinced, would have to avoid this fallacy by beginning with the given in experience and by analysing its most basic concepts. The fundamental task of metaphysics, as 'the science of the conceivability of experience', was to investigate the forms of experience, both metaphysically and psychologically (§93 Anm; VII, 159). It would ask a) how these forms arise (the psychological investigation) and b) what validity they should have (the metaphysical explanation).

There can be no doubt whatsoever, Herbart acknowledges, that Kant's philosophy marked a revolution in modern metaphysics. Kant not only saw through the fallacy of the old metaphysics – its conflation of essence and existence – but he also rightly stressed the need for metaphysics to remain within the limits of experience (§70; VII, 119). The starting point of all future metaphysics, Herbart believes, is Kant's thesis that existence is not a property but simply the absolute positing of a thing. It was another great contribution of Kant, Herbart adds, when he distinguished metaphysics from ethics, theoretical from practical reason, 'is' from 'ought' (§39 Anm.1; VII, 77). But, as important as these criticisms were, Kant still blunted their full force by advancing other doctrines that fed suspicious metaphysical speculation. His doctrine of the unity of reason blurred the distinction between theoretical and practical reason; and the concept of organism of the third *Kritik* not only confused the normative with the factual but it also encouraged speculation about nature as a vast living being (§39 Anm.1; VII, 77). Chiefly because of the third *Kritik*, Herbart claims, Kant thwarted his own reform of metaphysics.

Although Herbart thinks that Kant's philosophy was a great step forward, he also regrets that Kant did not come on the scene earlier when the old Leibnizian–Wolffian metaphysics was less decrepit (§99; VII, 176). That would have been enough to reform that metaphysics without destroying it. It could then have been built on the more solid foundation of existence and experience. Reclaiming that old metaphysics, and basing it on such a critical foundation, was Herbart's chief philosophical mission.

What was that old metaphysics? The monadology, of course. In the first chapter of the first volume of his *Allgemeine Metaphysik* Herbart sees the history of metaphysics since Kant as chiefly an attempt to reform Leibniz's monadology (§25 Anm; VII, 39). It is a far-fetched view of post-Kantian philosophy, perhaps, but it at least shows the importance that Herbart gave to Leibniz's doctrine. For him, the need to return to the monadology was clear and pressing: 'There is no need for a power of divination, not even for much true knowledge of metaphysics, to see that the monads, once their opponents die off, must come back to life once again' (§116; VII, 218).

What, specifically, did Herbart see in Leibniz's monadology? First and foremost, it was Leibniz's analysis of reality into simple self-sufficient units, entities which could serve as the ultimate elements or constituents of things. Like Leibniz, Herbart believed that there had to be such units; otherwise, analysis would proceed *ad infinitum* and there would be no reason to regard anything as real. Again following Leibniz, Herbart insisted that we could not understand these units in terms of extension, mass or impenetrability (Anm. VII, 43). These physical dimensions arose from the composition or aggregation of these units, so that they could not be attributed to any unit alone or in its intrinsic nature. On these grounds, Herbart argued for something like a Leibnizian distinction between reality and appearance: reality consists in simple real beings, and appearance consists in their composition or aggregation, which is the work of the perceiver. Again no less than Leibniz, Herbart understood his units as animate beings, as having a power of life (§374; VIII, 284–5). Hence he attributes to all real things, all ultimate units of reality, what he calls 'self-preservation' (*Selbsterhaltung*), that is, the power to resist other things intruding on them (§244; VIII, 117–18). This does not mean, however, that all monads have a power of representation. While all representation presupposes self-preservation, not all self-preservation takes the form of representation (see Herbart, *Lehrbuch zur Einleitung in die Philosophie*, §131; IV, 225). Because he saw self-preservation as the fundamental force of every being, Herbart criticized Kant for assuming that repulsion is a basic force behind matter. In postulating this force Kant took impenetrability as a datum, though it too stands in need of explanation (§154; VII, 299). Although Kant was right in criticizing the old view of matter as dead extension (§150; VII, 293), he did not go far enough, and he should have questioned the fact of impenetrability. Behind that impenetrability, and behind that force of repulsion, there lies another deeper power: self-preservation, the drive of everything to preserve or maintain itself, to repel forces impacting upon it.

So Herbart has great debts to Leibniz's monadology. But he also makes two criticisms of it, objections so basic that it is necessary to distinguish his monadology from that of his great predecessor. First, Herbart does not accept Leibniz's teleology, more specifically his assumption that a monad is a *nisus* or entelechy, a force that strives for self-realization (§132; VII, 262). While Herbart, in attributing self-preservation to real things, endows them with a primitive kind of life, he still wants that life to be understood on a mechanical basis. Second, Herbart maintains that there can, and indeed must be, interactions between monads, so that they stand in causal relations with one another; in other words, his monads have 'windows'.[9] While the intrinsic nature of a monad remains forever the same, impervious to all change, its extrinsic nature or properties depend on its relations to other things (§214; VIII, 76). Herbart thinks that Leibniz was perfectly correct to reject the doctrine of interaction as it was understood in his own day: an 'influx' where one body somehow gives its properties to another, as if properties could be somehow free-floating and detach themselves from one thing and attach themselves to another (§116; VII, 218). But Herbart maintains that we can have interaction without influx, provided that we make the right distinction. Namely, we must distinguish between *real* and *apparent* causality, where real causality is independent of time and place, and where apparent causality provides only a rule for succession for events in time (§146; VII, 287). That we must attribute real causality to substances Herbart leaves no doubt; he makes it a basic maxim 'No substantiality without causality!' (§220; VIII, 83).[10] Herbart had a strong motivation for attributing interaction to monads: it was only on that basis that he could explain the latest findings of chemistry. He is very much impressed with the advances of this new science, and he insists that metaphysics must come to terms with them (§378; VIII, 287) Accordingly, the final section of his *Allgemeine Metaphysik* is devoted to the explanation of chemical affinities on the basis of his monadology. To explain them, Herbart attributes to his monads relations of attraction and repulsion or what he calls 'mutual interpenetration'. Clearly, we are far removed from Leibniz's self-enclosed and self-absorbed monads.

---

[9] In his admirable summary of Herbart's philosophy, Röd maintains that Herbart's monads have no windows. See volume IX/1 of *Geschichte der Philosophie*, 188. However, Röd does not note the important distinction between real and apparent causality. He is also forced to recognize (188) that Herbart does need to postulate some form of interaction between his monads.

[10] Herbart's emphasis. This maxim seems difficult to understand in view of Herbart's critique of causal interaction in his *Lehrbuch* zur *Einleitung in die Philosophie*. See §106; IV, 164–6. But Herbart's arguments against the reality of interaction here should be understood as a critique of the influx theory.

3. OUTLINES OF MONADOLOGY

Now that we have some idea of what metaphysics should *not* be according to Herbart, we should proceed to examine what it should be. Let us attempt to reconstruct, in its barest outlines and from the ground up, Herbart's monadology. The foundational part of Herbart's metaphysics is what he calls its 'ontology', which appears in the first section of the second volume of his *Allgemeine Metaphysik*. We will not provide a commentary on this section – the task for an entire volume – but only pull together those ideas central to its foundation.

Herbart's ontology begins with some reflections on the starting point and method of metaphysics. It was the mistake of classical metaphysics, we have seen, to have begun with the concept of possibility, and to have neglected the connection of its concepts with reality (§166; VIII, 15–16). To rectify this mistake, to avoid constructing castles in the air, metaphysics must begin, Herbart demands, with reality itself, i.e. with the facts of experience, with what is given in sense perception (§161; VIII, 11). The proper method of metaphysics, just like that of natural science, is both analytic and synthetic (§164; VIII, 13). It must first analyse its facts, determining their basic components and their relations; it must then synthesize these components, putting them together again in their proper relations. It will then confirm its analysis if its synthesis reconstitutes the facts from which it began. Proper method in metaphysics is shaped like an arc, Herbart says. It begins with the given; it descends to the depths to explain the reality behind the given; and finally it returns back to the given (§164; VIII, 14). We shall hold Herbart's metaphysics to the demands of his own methodology; we will raise the question: Can it really explain the facts of experience?

Because metaphysics must cling to reality, Herbart insists that it has to begin with an analysis of the concept of reality itself (§199; VIII, 44). We must determine, in other words, what it means to ascribe reality to something. We are forced to do this, Herbart explains, because the sceptic gives us good reasons to doubt that the given – what appears to us in sensation – is real. Appearance and reality are the two poles of metaphysics (§193; VIII, 47). If what appeared were reality pure and simple, there would be no need at all for metaphysics. If we are to find reality, we must go beyond and get behind appearances; yet we still have to begin with appearances. As Herbart summarizes the predicament of the metaphysician: 'We doubt the reality of the given; we seek being; and our only hope of finding it still depends on the given' (§198; VIII, 52).

What, then, do we mean by reality? We have already seen that reality or existence does not add anything to the content of a concept (§202; VIII, 56). To say that something exists is not to attribute a new property to it. What exists first appears to us in experience as a thing with its properties. What we sense are the properties of something, for example, I sense that this apple is red, round, sweet and hard; but the thing itself, what unifies these properties, is not given in experience (§201; VIII, 56). These sensed properties presuppose some reality – the thing itself – of which they are the appearances (§199; VIII, 53). Reality then seems to be something in which properties inhere.

When we say that this thing is real, Herbart maintains, we assume that it has an independent existence. We assume that its existence does not depend on the existence of any other thing (§204, VIII, 59–61; §244, VIII, 177). If it were to depend on the existence of something else, it would not be entirely or completely real by itself, but it would be real only conditionally or hypothetically. If something is to be unconditionally or absolutely real, we must 'absolutely posit' its existence, so that its existence is 'full and complete' (§72; VII, 122). Because each fully real and complete being has such an independent existence, it is entirely indifferent to the existence of other things; it could just as well exist without these things; their existence is completely contingent for it (§244; VIII, 177). The absolute positing of a thing concerns its substance, what exists prior to its properties or accidents (§73; VII, 125).

This substance or real being, Herbart further insists, is absolutely simple (§73, VII, 124–5; §207, VIII, 63). It is *one* thing, completely integral and indivisible. For Herbart, like Leibniz, the very concept of existence involves that of unity. Leibniz made it an axiom: 'That what is not truly *one* being is not truly one *being* either' (Leibniz to Arnauld, April 1687, in *Die philosophischen Schriften*, II, 97). With that axiom Herbart fully agrees. It follows from it, he thinks, that we cannot take away anything from, or add anything to this unity, because addition and subtraction involve composition or decomposition, the assumption that ultimate reality has parts (§73; VII, 124).

At this point, we do well to ask ourselves: why does Herbart assume that there is a plurality of real beings? He points out that the concept of being itself is neutral about number, and that there is no reason from it alone to postulate one or many things (§208; VIII, 66). Yet the existence of many things is a fundamental premise of any monadology, and it is a premise that Herbart shares with Leibniz. Why? Part of the answer lies in Herbart's starting point: the given of ordinary experience. What is given to us in experience, he notes, are diverse complexes of characteristics, where each complex appears as one thing (§165; VIII, 15). These complexes are not free-floating or arbitrarily interchangeable because they appear together in our experience independent of our will and imagination (§169, VIII, 19–20; §171, VIII, 21). They therefore appear to inhere in something which unites them and makes them just one thing. It seems, then, that we should posit the existence of one substance or one thing for each complex. This is part of Herbart's reasoning, which is explicit in this passage from his *Einleitung in die Philosophie*:

> From the truly one, said Lucretius, there will never be many; from the truly many there will never be one. But many is given; therefore one must presuppose an original manyness.
>
> (§119; IV, 194)

But this answer only goes so far, because we could demand to know why these apparently many things are not just modes of a single universal substance. What makes us attribute independent reality to them? Herbart's answer to this question is apparent from his reaction against Spinoza's monism, which appears in the first volume of the *Allgemeine Metaphysik*. There Herbart makes it clear that there can be no transition from Spinoza's single substance to the modes of ordinary experience (§§46–48, VII, 51; §89, VII, 97). There is an unbridgeable gulf between the unity and eternity of substance on the one hand, and the multiplicity and temporality of its modes on the other hand. We cannot have, Herbart argued, both a one and a many: if the one becomes many, it divides itself and ceases to be one (§110; VII, 200). And so, if we are to explain appearances, the apparent reality of a multitude of things, we have no recourse but to embrace pluralism. We can do so, furthermore, with no need to worry about limiting the reality of substance. Spinoza feared that if there were two substances, then one would lose its reality by being limited by another. But this is a groundless worry, Herbart argues, because no substance loses any of its reality because of another (§73; VII, 125).

So far, Herbart has come as far as Leibniz in the first paragraphs of his *Monadologie*. He has postulated the existence of a plurality of simple substances, each of which is independent in essence and existence. All other reality is made of complexes of these simple substances. From the same basis as Leibniz, Herbart then draws very similar conclusions: that his simple substances are non-spatial and eternal. They are non-spatial because space is a form of composition, which comes from adding real things together (§265; VIII, 155). They are also eternal because time involves change, which means adding to or subtracting something from the reality of a thing (§281; VIII, 172). Space and time are therefore forms of the *appearance* of things, Herbart argues, because their reality depends on composition and quantity. In general, Herbart argues that we cannot attribute *quantity* to real beings because that would imply that they are divisible (§78, VII, 122;§208, VIII, 66). All quantities, all forms of aggregation or composition, concern only the *relations* between things but not the intrinsic reality of things themselves.

What about qualities? Do they not also involve some kind of division of the ultimate units of reality? We call our apple red, round, sweet and firm, where all these qualities are separable from one another. Indeed, they are so distinct that it seems impossible to refer to a single thing that unites them. Herbart fully recognizes this point, and insists that his ultimate units are an indivisible unity. When we see a thing as having different properties, he maintains, that is the work of our perceiving consciousness, which makes these distinctions to make sense of its experience (§213; VIII, 75). The utter unity and simplicity of substances does not mean, however, that they are bare substrata, completely bereft of any qualitative dimension. They have qualities; it is just that they are all bound together in an indistinguishable unity or whole.[11]

In accounting for the inherence of properties in things, Herbart develops one of his most distinctive and peculiar concepts: 'accidental views' (*zufällige Ansichten*). An accidental view of a thing arises from the perceiving consciousness, which attributes properties to the thing from its various perspective upon it. Herbart likens these accidental views to the solutions of a mathematical problem (§190; VIII, 45). They are similar to the various numbers which, when added together, give a certain sum; given these numbers, the sum is necessary, the only possible result. There are, however, many different numbers that could result in the same sum; and hence the view is 'accidental', because it is not the only set of numbers that gives the sum. There are as many accidental views of a substance as there are sets of numbers that could give a definite sum.

Ultimately, Herbart faces the same fundamental problem as Leibniz: How do we construct matter, space and time from the reality of simple things? The problem is pressing because these simple things are spaceless and timeless, so that adding them together, forming complexes from them, should not result in space and time. Herbart's solution to this problem is similar to Leibniz's: he attributes the reality of space and time to the perceiver. The forms of composition arise from the consciousness of the subject, which joins together the multitude of independent things so that they appear in space and time (§265; VIII, 154). Herbart insists that neither space nor time, neither motion nor change, are properties of things themselves; they are only appearances of things for some consciousness. They belong to what Herbart calls the sphere of 'objective appearance', where an objective appearance consists in how *any* intelligent being *must* perceive a plurality of independent real things (§293; VIII, 187). An objective appearance differs from a subjective one in that it *must* hold always and for all perceivers, whereas a subjective appearance holds only sometimes for some perceivers. What Herbart means by objective appearance is pretty much what Leibniz calls a 'phaenomena bene fundata', i.e. an appearance of a thing based on its reality, though it is not a picture or perfect image of a thing.

To give some idea of the difficulty Herbart faces in reconstructing appearances from real beings, it is instructive to consider his construction of matter in Part II of his *Allgemeine Metaphysik*. To explain matter we have to account for cohesion, the joining of simple entities to form a whole. For two entities to come together, they must interact. But to assume interaction requires that things somehow penetrate one another (§271; VIII, 161). There must be, as Herbart puts it, 'imperfect togetherness' or 'partial penetration'

---

[11]This is the point Herbart would make against Cassirer's criticism, *Das Erkenntnisproblem*, III, 401, that his real beings are propertlyless and therefore 'the *caput mortuum* of abstraction'. His reals have characteristics or properties; it is just that they are dissolved in a unity, and that we have to grasp them separately and therefore not as they are in themselves.

of one thing by another. There must be *penetration* because if things were entirely impenetrable, they would not attach to one another; nothing would aggregate. There must also be *partial* penetration, because if they were *entirely* penetrable then one thing would become coincident with the other; they would be congruent and one and the same thing (§267; VIII, 158–9). But, though it is indispensable, Herbart admits that the concept of partial interpenetration is a fiction because it has to assume that simple entities are somehow divisible, that one part of them is penetrated but not the other. Real beings, however, do not have parts (§278; VIII, 167). Herbart then asks himself in a telling passage: Why assume imperfect togetherness, partial penetration, if it is only a fiction? His answer is disarming: we must assume it because real things form matter, and the existence of matter is a phenomenon that we must explain (§278; VIII, 167). The construction of matter therefore requires admitting a fiction, indeed a contradiction (i.e. that the indivisible is divisible), into our ontology.

The same problem arises for Herbart's construction of ideal space. Herbart distinguishes between *ideal* and *real* space: ideal space is that which arises from the interaction of monads themselves; real space is that which appears in our ordinary sense experience, and which arises from the aggregation of monads by our sense organs and consciousness (see *Lehrbuch zur Einleitung in die Philosophie* §134; IV, 235). Herbart's construction of ideal space, which appears in Chapters 2–4 of the third section of his *Allgemeine Metaphysik,* is one of the most complicated and subtle theories of the entire book. After the most exhausting and exacting account of how to form lines, planes and solids from the interaction of monads, Herbart finally arrives at a three-dimensional space akin to that of ordinary experience. Herbart is confident that, if we have the proper psychological explanation, we can ultimately identify this ideal space with the real space of our ordinary experience (§264; VIII, 152–3). Yet the derivation, for all its subtlety and sophistication, is a failure. Herbart's real space, like cohesion, presupposes the partial interpenetration of monads, and so it involves the same difficulties as that concept (§265; VIII, 155).

So, in the end, Herbart has as much difficulty in accounting for the realm of appearance – space, time and motion – as Spinoza. His metaphysics did not have the arc shape he demanded: it descended into the depths of the ultimately real; but never again did it ascend to the realm of appearance. The only way it could lift itself back to that level was by resorting to fictions, by dividing the indivisible and by connecting the unconnectible.

4. HERBART'S REALISM?

In his popular study of Herbart, Otto Flügel, a noted Herbart scholar, once wrote: 'He who knows only a little about Herbart knows that he was a realist at a time when his age mostly thought idealistically' (*Herbarts*

Lehren und Leben, 1). 'Realism' has indeed been the catchword to describe Herbart's philosophy. The signal doctrine of his philosophy, according to the textbooks, is its 'realism', which is supposed to be its distinguishing feature from all the idealist systems of his day (see e.g. Ueberweg, *Grundriss der Geschichte der Philosophie*, 107–27). There must be something to the interpretation, one would think, given that Herbart himself would sometimes refer to his philosophy as 'my realism' (Herbart to Brandis, 1 Oktober 1831, *Sämtliche Werke* VIII, 412).

Any attentive reader of the above paragraphs will be rightly puzzled by this interpretation, however. How, he or she will ask, can Herbart be a realist if he denies the reality of matter as an extended, solid thing? How, indeed, can he be described as such if he maintains that the world of space and time are only an appearance of the ultimately real for the conscious subject? Is not such a doctrine better described as a form of idealism?

There is indeed a problem here. The realist interpretation of Herbart's philosophy stands in need of drastic qualification. To give the reader a sense for the problem, consider the following lines from Herbart's *Einleitung in die Philosophie*:

> We are completely enclosed inside our concepts; and for just that reason, because we are so, our concepts decide the real nature of things. Whoever holds this to be idealism (from which it is completely different) must know that, according to his usage, there is no other system than idealism.
>
> (§114; IV 183)

Though Herbart himself does not adopt this usage, he still affirms the doctrine behind it, a doctrine that would usually be regarded as idealism. Why not, then, just call his philosophy 'idealism'? There is all the more reason for doing so, when we consider that there are other passages from Herbart's works where he endorses doctrines that can be described only as 'idealism'. Thus he maintains explicitly, as Kant once had, that the immediate objects of our knowledge are only representations, that we construct an external world from them, and that we cannot get outside our representations to know things-in-themselves (see *Lehrbuch zur Einleitung in die Philosophie*, §103; IV, 159–60; and *Hauptpuncte der Metaphysik* §3, II, 192). Herbart thinks that these are fundamental truths of idealism, which no philosophy can or should bring into question.

All this poses the question: What, if anything, does 'idealism' and 'realism' mean in Herbart's philosophy? In what sense is his philosophy realist, and in what sense is it idealist?

As much as Herbart affirms a kind of idealism, he also sometimes distances himself from it, and he even talks about the need for 'a refutation of idealism'. It is just such passages that seem to lend fuel to the realist interpretation of his philosophy. What exactly Herbart meant by 'idealism' becomes clear from section four of his *Allgemeine Metaphysik*, which is devoted expressly to exposing the deficiencies of idealism. It becomes clear from Herbart's exposition that 'idealism' has a very specific meaning. It refers to the idealism of Fichte's *Wissenschaftslehre*, which Herbart, following Jacobi, regarded as the purest and most radical version of idealism (see *Allgemeine* Metaphysik, §141; VII, 276. See also *Lehrbuch zur Einleitung in die Philosophie*, §104; IV, 162. See the 'Anmerkung' to the second, third and fourth editions). Fichte's idealism represents the boldest form of idealism because it sees everything as the product of the self-positing ego. Herbart argues, however, that such an idealism is implausible for the simple reason that the purely self-positing ego cannot explain the reality of the non-ego (*Allgemeine Metaphysik*, §§323–6, VIII, 228–33). The ego is held to be perfectly self-positing; but it also posits a non-ego, which is opposed to itself. How does that which is purely self-positing limit itself and posit something opposed to itself? How do we derive from a purely self-positing ego something self-opposing? We cannot, Herbart complains, and not least for this reason we should abandon pure idealism.

But this is only the beginning of the problems for idealism, according to Herbart. He further argues that not only the non-ego in general, but also the variety of qualities given in our experience – the so-called manifold of experience – cannot be derived from the ego. What sense qualities appear to us, and when and how they appear, is strictly contingent for the purely formal activities of the absolute ego. Herbart stresses that it is not only the *matter* of experience that is given, i.e. the simple existence and quality of sensations, but also its *form*, i.e. in what order these sensations appear to us (*Allgemeine Metaphysik*, §118, VII, 223 and §201, VIII, 55). The specific properties that appear to our senses are simply given to us, and they occur in just this order rather than another, completely independent of our will and imagination. When we see our apple as red, round, hard and sweet, for example, we cannot substitute these properties with any others; precisely what we see, and where, when and how we see it, depends on factors beyond conscious control. What we see is partly the result of things-in-themselves, which exist independent of our consciousness, and partly the result of perceptive organs and activities, so that what we see are appearances of things-in-themselves.

It should be clear from these arguments that we cannot describe Herbart's philosophy as an unqualified idealism. They show that Herbart thinks that idealism, in its pure Fichtean version, is an utterly unworkable and implausible philosophy at odds with the facts of experience. The facts of experience are the great stumbling block of idealism, because they show that not only the matter but also the form of experience is given to us.

But it should also be clear that Herbart's philosophy also cannot be described as an unqualified realism. Herbart is not a *direct* realist, i.e. someone who holds that we have an immediate knowledge of the external world, because he does not think that our representations directly reflect, copy or mirror given characteristics; all that we are directly aware of, he insists, are our own representations. But Herbart is also not an *indirect* realist, that is, someone who thinks that we can *infer* the characteristics of things-in-themselves from our representations. There are two problems with this position: first, that the nature of our representations depends so much on the psychological and physiological apparatus with which we perceive things that it becomes impossible to separate out what does and does not represent reality itself; and, second, that reality in itself is very different from the reality that we perceive – it is eternal and spaceless, whereas what we perceive is in space and time. Finally, Herbart is also not a *scientific* realist, that is, someone who assumes that the formal, quantifiable or mathematical properties of things are their real properties; for, as we have seen, he denies that reality in itself is quantifiable and that it exists in space and time. For Herbart, the quantifiable aspects of a thing are indeed its *least real* aspects.

There is still, however, an important sense in which Herbart's philosophy can be described as realist after all: Herbart thinks that there is a realm of things-in-themselves, and that these things-in-themselves exist apart from and prior to our consciousness or perception of them. Although we cannot ascribe to things-in-themselves, as they exist apart from and prior to our experience, the sense properties that we know from experience, we are still justified in regarding these properties as *appearances* of things-in-themselves, that is, as how they appear to human beings with their sensibility and forms of understanding. These appearances are, in part, the effects of how things-in-themselves act upon us. In this rather minimal sense Herbart is no more a realist than Kant, who also expressly maintains the reality of things-in-themselves and who protests the conflation of his philosophy with idealism on just these grounds. Still, we can understand Herbart's insistence on the term 'realism' when we remember Kant's own difficulty in distinguishing his transcendental idealism from Berkeley's idealism. In a revealing passage from his *Hauptpuncte der Metaphysik* Herbart claims that Kant made a mistake in calling his doctrines of the nullity of space, time and motion, and ignorance of things-in-themselves, a form of idealism (§10; II, 205). These doctrines are a form of realism, in his view, because they still presuppose the existence of things-in-themselves.

If we strictly define Kant's transcendental idealism in its intended sense – as a the doctrine of the distinction between appearances and things-in-themselves – then Herbart's own philosophy is very close to Kant's. In a remarkable passage from his *Allgemeine Metaphysik* Herbart virtually admits as much, stating that his doctrine can be described as 'idealist' and that it is closer to Kant's views more than anyone else (§298; VII, 193). In one respect, though, Herbart departs from Kant in providing a more unequivocal basis for his realism: he thinks that we are perfectly justified in extending the categories beyond sense experience; he denies that the principle of causality has only a temporal meaning, as if it held only for succession in time (see *Lehrbuch zur Einleitung in die Philosophie*, §115; II, 184. See 'Anmerkung 2' to the second, third and fourth editions. Also see *Allgemeine Metaphysik*,

§76, VII 126; §146 and VII 287; and §299, VIII 193). Hence it is possible to infer the existence of things-in-themselves and that they are the causes of our sensations.

The heart of Herbart's realism *and* idealism appears in his concept of 'objective appearance', which he expounds in his *Allgemeine Metaphysik* (§§202–3; VIII, 186–8). The realm of objective appearance appears in space and time, where space and time are appearances because they depend on the perceiving subject, but where they are also objective because they are universal and necessary characteristics of these appearances. Although Herbart claims that his doctrine is akin to Kant's idealism more than anyone else, it would have been more accurate for him to have said that it is closer to Leibniz. For he holds, just as Leibniz once had, that the appearances of things in space and time are due to the effects of things-in-themselves on our sensibility, and that space and time are confused representations of things-in-themselves (see *Einleitung in die Philosophie*, §95; IV, 147. See the 'Anmerkung' to the second, third and fourth editions). Appearances are indeed for Herbart, just as they were for Leibniz, necessarily tied to things-in-themselves, so that they are attached to them as appearances *of* things-in-themselves; they cannot be detached from them as if they were nothing more than representations in consciousness. What we perceive in our experience is 'an accidental view' of the thing-in-itself, i.e. it is an aspect or property of that thing as it appears to our perceiving consciousness. It is indeed telling that Herbart objected to Kant's own distinction between appearances and things-in-themselves on the grounds that it left a too sharp separation between them (*Allgemeine Metaphysik*, §299; VIII, 195). Here Herbart targets Kant's frequent references to appearances as mere representations in us.

All in all, then, it is a rather slender and minimal sense of 'realism' that Herbart advocates. It amounts chiefly to two claims: (1) that there exist things-in-themselves apart from and prior to consciousness; and (2) that experience consists in appearances of them. Still, we can understand why Herbart would want to call this doctrine realism. It was an important distinguishing characteristic of his philosophy in contrast to Fichte's, which was seen as the most consistent and radical form of idealism in his day. Herbart, who had once been a Fichtean idealist, was very keen that his philosophy not be confused with the idealism of his former mentor. What better way to keep a distance from him than by calling his philosophy realism? Yet, for all the reasons we have seen above, the epithet become very misleading, disguising Herbart's own fundamental idealist doctrines.

It is significant that Herbart had classified Leibniz as a realist (*Allgemeine Metaphysik*, §115; 211). He was a realist for the same reasons as Herbart. He too affirmed the reality of things-in-themselves; he too claimed that we perceive appearances of them; and he too claimed that they are the basis for our perception of an external world. All the more reason, then, that Herbart would want to identify with Leibniz. This was one final affinity of his philosophy with that of the great rationalist, whose doctrines he had done so much to rehabilitate.

BIBLIOGRAPHY

Bréhier, E. *The Nineteenth Century: Period of Systems 1800–1850.* Translated by Wade Baskin, 111–203. Vol. 6 of *The History of Philosophy.* Chicago: University of Chicago Press, 1968.
Cassirer, E. *Das Erkenntnisproblem in der Philosophie und Wissenschaft der neueren Zeit. Die nachkantische Systeme.* Berlin: Cassirer, III, 378–410.
Copleston, F. *Fichte to Hegel.* Vol. 7 of *A History of Philosophy.* New York: Image Books, 1965.
Erdmann, J. *Die Entwicklung der deutschen Spekulation seit Kant.* Vols 5–6 of *Versuch einer wissenschaftlichen Darstellung der Geschichte der neuern Philosophie.* Stuttgart: Frommann-Holzboog, 1977.
Fischer, K. *Geschichte der neueren Philosophie.* Vols. 5–7. Heidelberg: Winter, 1872–77.
Flügel, O. *Herbarts Lehren und Leben.* Zweite Auflage. Leipzig: B.G. Teubner, 1912.
Hegel, G. W. F. *Werke.* Edited by Karl Michel and Eva Moldenhauer. Frankfurt: Suhrkamp, 1971.
Herbart, J. F. *Hauptpuncte der Metaphysik.* Göttingen: J.C. Baier, 1806.
Herbart, J. F. *Sämtliche Werke.* Edited by Karl Kehrbach and Otto Flügel. 19 vols. Langensalza: Hermann Beyer & Söhne, 1887–1912. Reprinted: Aalen, Scientia Verlag, 1989.
Kant, I. *Schriften.* Edited by Prussian Academy of Sciences. Berlin: de Gruyter, 1902.
Kroner, R. *Von Kant bis Hegel.* Tübingen: Mohr, 1921.
Leibniz, G. W. *Die philosophischen Schriften von Gottfried Wilhelm Leibniz.* Edited by C. J. Gerhardt. Berlin: Weidmann, 1875–90.
Röd, W. *Geschichte der Philosophie.* Munich: Beck, 2006.
Rosenkranz, K. *Geschichte der kant'schen Philosophie.* Leipzig: Voss, 1840.
Sandkühler, H. J., ed. *Handbuch Deutscher Idealismus.* Stuttgart: Metzler, 2005.
Schmitz, J. N. *Herbart Bibliographie 1842–1963.* Winheim: Julius Beltz, 1964.
Solomon, R. C. *Continental Philosophy since 1750: The Rise and Fall of the Self.* Oxford: Oxford University Press, 1988.
Spinoza, B. *Opera·Werke.* Darmstadt: Wissenschaftlicher Buchgesellschaft, 2008.
Ueberweg, F. *Grundriss der Geschichte der Philosophie.* Vierter Theil. *Das Neunzehnte Jahrhundert*, Neunte Auflage, edited by Max Heinze. Berlin: Mittler und Sohn, 1902.

Article

Bolzano's Monadology

Peter Simons

> Bernard Bolzano (1781–1848), known in his lifetime as 'the Bohemian Leibniz', is best known as a logician and mathematician, but he also developed a monadology in which the monads, which he called 'atoms', have spatial location and physical properties. This essay summarizes and assesses his monadology.

Bernard Bolzano (1781–1848) is best known for the throroughgoing and astonishingly modern semantic reform of logic proposed in his monumental *Wissenschaftslehre* (1837). While little known or regarded in his time or for a long period thereafter, this has gradually come to be recognized as one of the most substantial contributions to the subject between Leibniz and Frege. He is also known as an innovative mathematician, an early proponent of clear, purely mathematical conceptual foundations for the subject, an enthusiastic supporter of the actual infinite, and a forerunner of Cantor in set theory. Much less known is Bolzano's metaphysics, which is a thoroughgoing monadology, proposing the existence of an infinity of partless substances, physical and mental, completely filling space, and interacting via forces of attraction and repulsion. Bolzano closely studied the critical philosophy of Kant, and disagreed with it on many points. His advocacy of the actual infinite and his confidence in the idea of a realist cosmology of simple substances represented a qualified return to Leibniz, whose work he admired greatly, but with aspects of which he also disagreed. Like Leibniz, Bolzano had polymathic talents – not just logic and mathematics, but also theology, ethics, aesthetics and physics were all enriched by his work – and he was known even in his own times as 'the Bohemian Leibniz'. Bolzano was one of the nineteenth century's clearest and most systematic thinkers, so his metaphysics deserves to be better known: this essay attempts a critical summary. Yet his views particularly as regards their import for physics are also irremediably archaic, coming just before the electromagnetic revolution of Ørhsted, Ampère, Faraday and Maxwell, remaining wholly descriptive rather than quantitative in form, and in Bolzano's striving for apriority, beyond verification or falsification. The failure of a mechanical substance–attribute monadology of even such an astute thinker as Bolzano must raise doubts as to whether any such account is viable.

SOURCES

Bolzano's mature metaphysics is found in three main sources, given here in chronological order. The first is his large monograph *Athanasia*, first published anonymously in 1827, then in an enlarged and improved second edition in 1838 under the author's name. The only work of Bolzano's to achieve any popularity in his lifetime, it is, as the title says, a defence of personal immortality. Its full title in approximate English translation is *Athanasia, or Reasons for the Immortality of the Soul. A Book for Every Educated Person Who Wishes to be Reassured about This*. The second source is Bolzano's 'Aphorismen zur Physik' ('Aphorisms on Physics'), dating from 1840, read and discussed in 1841 at the Royal Bohemian Society of Sciences, but never put into a form he deemed worthy of publication. It was edited and published in the *Gesamtausgabe* (Complete Edition) by Jan Berg in 1978. The final major source is the second half of *Paradoxien des Unendlichen* (*Paradoxes of the Infinite*) of 1851, edited by and published posthumously in 1851 by Bolzano's pupil Franz Příhonský. Bolzano worked on the *Paradoxien* in 1846–7. Its title is apt to mislead. It is in fact a sustained *defence* of the notion of infinity against claims that it is paradoxical, absurd or self-contradictory. The first part of the book deals with infinity in mathematics, but the second part, which is what interests us here, deals with infinity in physics, or in the physical world. A more compact but less complete source of information on Bolzano's monadology is an essay, 'Atomenlehre des sel. Bolzano' ('Atom Theory of the late Bolzano') by Příhonský. While Bolzano did add a little to his views over time, he revised relatively little, so we can treat his monadology as a theoretical unit and do not need to distinguish different phases of development.[1]

SUBSTANCE, ADHERENCE AND ATTRIBUTE

As argued in *Athanasia* and presupposed elsewhere in Bolzano's writings, he accepts that among real things (*wirkliche Dinge*), those which are to be

---

[1] In this essay we shall cite *Athanasia*, 'Aphorismen zur Physik' and *Paradoxien des Unendlichen* using the respective abbreviations *AA*, AP and *PdU*. References to *AA* are to page numbers of the second edition; references to AP and *PdU* are to Section numbers to enable any edition or translation to be used. Other abbreviations of occasionally cited works are given in the references and are also to sections unless otherwise indicated.

found in space and time, and enter into causal relations, there are two fundamental sorts: substances and adherences. A substance is an ontologically self-sufficient thing that 'exists for itself' (*AA* 21), is a non-predicable (*AA* 1) individual (*WL* 142, *PdU* 57), as in the tradition from Aristotle to Leibniz. Adherences are the individual characteristics of substances and of other adherences (*AA* 9). Adherences are *of* or *in* other things (*an etwas Anderem*), substances are not – for example the matter of which a body is composed is substance (*AA* 21). Adherences are particulars which depend on their substances for their existence and which characterize or qualify these substances, and can be called their properties (*Eigenschaften*) (*WL* 272.2). For example the colour, weight and smell of a body are adherences of it (*AA* 21). Although not always quite consistent in his usage, he tends to call the kinds or types of adherences 'attributes' (*Beschaffenheiten*), but like Leibniz he thinks the real things, things having existence (*Daseyn*) are all particular (*AA* 9). Though no nominalist – unlike Leibniz – Bolzano does not make great play with the notion of universal or general attributes in regard to real things. He does allow that non-real things such as concepts and numbers have attributes (*AA* 22), and cannot have adherences, since the latter are real. For present purposes it is enough to say there are substances, and how they are is determined by what adherences they have. The spatial location of a substance – since for Bolzano, unlike Leibniz, all substances are in space and have physical characteristics – is not an attribute or adherence but what Bolzano calls a determination (*Bestimmung*) (cf. Morscher, *Das logische*, 69 ff.). Likewise, the time at which a substance has a particular adherence is a determination of the substance: that which has the attribute 'is sitting' because of the existence of an adherence of Socrates's sitting is not Socrates *simpliciter* but Socrates at such and such a time. '[B]y the word "time"', writes Bolzano, 'we mean nothing but that particular determination of an actual thing which is the condition for ascribing truly a certain attribute to that thing' (*WL* 79). Neither space nor time is real, since they are neither substances nor adherences (*PdU* 17), nor do they have effects (*WL* 79).

THAT THERE ARE SUBSTANCES

Bolzano has an argument for the existence of substances (*AA* 22) which is a close analogue of his cosmological argument for the existence of God. It goes as follows.

(1) Something real exists.
(2) Everything real is either a substance or an adherence.
(3) There either are substances or there are not.
(4) If there are not, then everything real is an adherence.
(5) But adherences are all dependent entities.

(6) Therefore the totality (complete collection) of adherences is also dependent.
(7) It must then depend on something real that is not an adherence.
(8) This must then be a substance.
(9) Therefore at least one substance exists.

The flaw in this argument – leaving the classificatory presumption 2 aside – is step 6. Why could not all adherences be dependent – that is, each be dependent on *something* – yet groups of them which depend mutually on one another not constitute substances? The analogous flaw is present in the cosmological argument (Ganthaler and Simons, 'Bernard Bolzanos kosmologischer').

Bolzano notes that since we know we have ever changing ideas and judgements, since these changes are real, and we can be certain that they must be adherences, their substances are the bearers of change, so that the mental and physical properties and events that affect us are our adherences, but we ourselves are substances (AA 25).

ATOMS

Bolzano makes a fundamental distinction among real things between those which do and those which do not have (proper) parts. Real things with parts are composed (*zusammengesetzt*) of these parts, and Bolzano calls them collections or aggregates (*Inbegriffe*). A collection is simply anything which is composite, that is, which has (proper) parts. Bolzano does not make a clear distinction between collections as pluralities of individuals (e.g. the collection of books in my study) and collections as composite individuals (e.g. the single clock in my study), and it is futile to look for such a distinction – later made much clearer by such writers as Russell and Leśniewski – in his work. All composite objects (collections) are made of simpler objects, and Bolzano holds that it is obvious they are all ultimately composed of objects which are utterly simple, so lacking parts. These correspond in their simplicity and fundamentality to Leibniz's monads, but Bolzano disagrees with Leibniz about the relationship of his simples to space and to one another. Whereas Leibniz's monads are strictly not in space and have no spatial characteristics, being all of the type of indivisible mental substances or souls, Bolzano, in keeping with common sense and the physics of his day, takes his atoms to have a spatial location or determination, which varies when the atom moves. Since atoms have no parts, they cannot spread out over a volume and therefore at any one time the location of an atom is a spatial point (*AA* 117, AP 4). Because atoms have no extension, they have never been perceived (*PdU* 50), but despite this, it is 'laughable' to suppose for this reason that they do not exist (ibid.) Bolzano holds that the crucial assumption leading Leibniz to his idealistic or panpsychistic conception of monads is the principle of pre-established harmony, which arises out of Leibniz's attempt to overcome the mind–body problem and to avoid the assumption of transeunt causation. Bolzano thinks he can happily and successfully explain both mind and body through the forces of attraction among his atoms, and so need not deny that there are causal interactions among atoms and so between mind and body. Bolzano's theory of simples is thus, like that of the pre-critical Kant, also a physical monadology, the physical and the mental being co-comprised among real substances and adherences. He seems not to have known or at least cited this early Kant work, but does know the not dissimilar work of Boscovich, whose forces Bolzano claims to be substances rather than adherences (*PdU* 57), whereas he takes his own adherences to be forces (*PdU* 57) or mental phenomena (*WL* 143), whose bearers are substances, ultimately atoms.

ATOMS ARE INDESTRUCTIBLE

Like Leibniz, Bolzano argues that atoms cannot be created or destroyed by natural means (*AA* 69), since the only way in which natural processes can create or destroy is by the rearrangement or reconfiguration of atoms, by their coming together and dispersing again. No atom has a beginning in time (*AA* 74) or an end (*RW* 214–27). Thus in the *Athanasia*, which has a deliberately consolatory message, and was written with Bolzano's friend Anna Hoffmann in mind, the death of whose daughter had first brought them into contact, those atoms which are human souls are likewise indestructible, and so exist forever (*AA* 68–84). Whether such persistence endows or re-endows them with the kind of conscious mental characteristics they enjoy during life is another matter, answerable less by natural philosophy than by religious faith. That there are atoms and among them souls or minds is taken by Bolzano to be evidenced by introspection, of mental phenomena together with the recognition that these themselves are adherences (*AA* 16, 25). God's creation of atoms, including ourselves, cannot therefore be creation by bringing into existence at a point in time, but must be the holistic, non-temporal creation of the atom as an object existing throughout time (*AA* 79).

INTERACTION OF ATOMS

The adherences of atoms are forces that act on other atoms (*PdU* 57). The forces can be attractive (*AA* 48) or repulsive (AP 19, *PdU* 63–4). (The thought that they might not act along the line between the atoms, as in electromagnetic induction, appears not to have occurred to Bolzano.) The German word *Kraft* that Bolzano uses here can mean both 'force' and 'power', and so covers both the disposition to act and the acting itelf. There is mutual interaction between atoms through their adherences.

In fact, every atom permanently and continuously influences every other (*PdU* 60), so there is instantaneous action at a distance (*AA* 67), but no interpenetration: no point can be occupied by more than one atom (*PdU* 54). All atoms are changing all the time (*PdU* 50), and there is no 'dead' or purely inert matter (*PdU* 51). The intensity of the interaction or mutual influence (*einwirken*) varies considerably; however, as not all atoms have the same power (*AA* 21), they move relatively to one another, and the density of material objects varies and is not uniform (*AA* 16). Bolzano's is a through and through causal monadology, unlike that of Leibniz. He readily accepts a Newtonian inverse square law for the mutual attraction of atoms (*AA* 27), but not at all distances. Noting that solid bodies are hard to compress, he postulates, in ideas not dissimilar to those of Boscovich, that for any two atoms there will be a distance below which they mutually repel, and a distance above which they mutually attract (AP 63, *PdU* 63). The distance g below which mutual repulsion operates may be smaller than the distance G above which attraction operates. Distances between endow the atoms with a 'comfort zone' in which neither attracts nor repels the other but as it were 'happy' or 'comfortable' with them: Bolzano uses the German words *genehm* ('acceptable') and *wohltuend* ('beneficial'). When an atom has another one closer to it than g, it is 'uncomfortable', meaning it will endeavour (*bestreben*) to repel the other, whereas atoms further away from it than G it 'prefers' to have closer and so will attract them (*PdU* 63). The size of a substance's comfort zone may vary with time (*AA* 21). Despite their standard psychological meanings, the words in Bolzano's usage are purely mechanistic in import. When two atoms are at a comfortable distance, that simply means they are disposed neither to attract nor to repel one another. This can be illustrated by the way in which magnetic levitation repulsive magnetic forces balance attractive gravitational forces and the bodies remain in equilibrium and at rest (not Bolzano's example).

DOMINANCE

Among atoms, some exercise a greater influence on those around them than any of those around them exercise on them in return. Bolzano calls such atoms *dominant* (*herrschend*) or distinguished (*ausgezeichnet*). A dominant atom influences the infinitely many atoms around it that have less power, though no dominant atom has infinitely more power than other atoms (*PdU* 62). The idea, though not its causal interpretation, is already present in Leibniz. Bolzano goes so far as to identify dominant substances as mental and dominated substances as material, though he admits that the difference is one of degree and not sharp. By taking the mental and the physical both to consist of atoms of variable dominance, he takes himself to have solved the mind–body problem: since the soul is an atom (indeed a dominant one) and the body consists of many atoms, there is no mystery about their interaction, which in some cases is immediate (*unmittelbar*) (*ΛΛ* 64, *PdU* 56). The soul, a spiritual (*geistige*) substance, is in space just as much as is the body (*PdU* 55). Body and soul develop together, and human reproduction shows that they are of the same kind (*gleichartig*) (*AA* 59). That Bolzano can hardly be taken to have *solved* the mind–body problem in such a way is not at issue, but there is no doubting the directness and boldness of his proposal.

MOTION AND OTHER CHANGE

Every atom is at any time at a point in space (*AA* 4) and a point cannot host more than one atom (*AA* 17). Bolzano's physical world is a plenum: there is no space that is not completely filled by atoms, not even a single point. But motion is possible even in a plenum (*AA* 15), because despite completely filling space, the atoms in it can be compressed together without any of them coinciding and expanded from one another without leaving holes. To motivate this thought, Bolzano notes, using a now-familiar argument, that there are just as many points on a line of given length as there are on a line of greater or lesser length (*AA* 17, *PdU* 20). The variability of density of the aether and other occupants of space allows for the transmission of waves by compression, rarefaction and lateral motion, about which Bolzano would latterly have been well informed through his friendship with Christian Doppler.

AETHER

Most of space is empty of bulky matter. The balance is made up by atoms of aether, which is a pervasive, space-filling fluid (AP 25) or world-stuff (*Weltstoff*). Evidence for its existence comes from the wave theory of light, delays of comets and the shape of comet tails. Radiant heat passes through the aether as a transmissive medium (AP 52). Its existence is unverifiable because it pervades all matter so its mass cannot be determined (AP 27). Loss of aether mass via radiation also cannot be tested because mass and volume are not in a fixed proportion: density varies (AP 16). Every atom attracts aether atoms, but atoms which attract other than just aether atoms are distinguished (AP 29). Aether atoms attract only one another so are the least dominant or most dominated of all; distinguished atoms are all at a level above aether atoms, but distinguished atoms also come in different kinds with different degrees of dominance.

BODIES

Material bodies are finite in volume whereas atoms have zero volume. Therefore, every material body is made up of infinitely many atoms (we would now say uncountably, infinitely many – Bolzano would probably have been delighted by Cantor's discovery of a hierarchy of infinities). Of these, almost all are aether atoms, and only a finite number are dominant. The dominant atoms in a body are what make matter detectably massive, forceful and resistant, even though they are only finite in number in a finite volume. The boundaries of bodies are not sharp in the sense that there is an abrupt end to the matter. Any dominant atom or collection of them drags a halo or shell (*Hülle*) of aether atoms around with it, and these can be loosely enough bound to transfer their allegiance from one body to another. For example, when Bolzano picks up his quill to write, some of the aether atoms bound to the quill more than to the surrounding air will be 'stolen' by (tend to move with) his fingers. The limits of a body are those aetherial atoms which are the furthest from it that are still attracted to it more than to other, neighbouring dominant atoms. So if the body moves, the aetherial atoms will go with it, unless they are 'stolen' by some other more attractive neighbour (*PdU* 66–7). The boundary of a body will be always changing. What we commonly regard as the touching of bodies is simply their coming into close proximity and their outlying dominant atoms getting too close for comfort (closer than g) and so repelling one another. The intervening space is at all times filled with aether, but that may get compressed in volume.

Bodies made up of different kinds of materials have different natural densities, as is proved by weight measurements of objects of the same volume but different masses, or the same mass but different volumes, it being assumed that weight and mass are always proportional, and weight arises through the mutual attraction of a body and the earth (*PdU* 63). Mass and density are not simply a measure of the number of atoms in a volume, since all volumes contain the same number of atoms, and a volume of 'empty space', that is, containing just aether, has a density far below that of one filled by a body.

As revealed by chemistry, bodies consist of a number of elements or basic stuffs (*Grundstoffe*). If an atom attracts atoms of kind A more than other kinds, it is said to have an affinity (*Wahlverwandtschaft*) with the kind A (AP 29). A basic stuff consists (apart from aether) of atoms of a single kind which have an affinity for one another (AP 36). It is affinities that explain chemical reactions (AP 37).

CRITICISM

Bolzano's monadology, which we have here summarized in very compressed form, is comprehensive and ingenious, and represents something like the limit to which one can take a monadology in accounting for the dynamical phenomena of classical physics. Bolzano was, through his own reading and his acquaintance with leading scientists of his day such as Doppler, well informed on the state of contemporary science. That subsequent developments in physics and chemistry radically undercut many of the assumptions underlying Bolzano's account is not his fault, though the extent to which he is prepared to treat his account as a-priori metaphysics rather than fallible science is a measure of his questionable rationalism. While characteristic of its time, his assumption of an aether takes on a disturbing tinge of apriorism, in that he regards the lack of direct evidence as insignificant by comparison with non-empirical arguments for the aether. It would have been interesting to know how he would have reacted to the Michelson–Morley null results which indicate the absence of detectable motion of the earth through an aether. Would he have accepted the evidence at face value, or would he have sought an explanation preserving the aether, for example, that the earth drags a large, measurement-neutralizing halo of aether around with it?

In his prefatory notes to AP, Jan Berg summarizes and criticizes Bolzano's metaphysics by noting that it

> attempts to explain almost all natural phenomena unitarily and at one go. For this Bolzano needs on the one hand a minimum of very general basic concepts, but on the other hand infinitely many special laws of motion. This dynamic contradiction in Bolzano's metaphysical system seems to imply that no verifiable predictions can be made on its basis.
>
> (*BGA* 2 A 12,3, 112, my translation)

Since it is by metaphysical design good for everything, it is scientifically good for nothing. This empirical insufficiency, rather than echoes of the official persecution of Bolzano, probably explains adequately enough why there is a clear discrepancy in influence between the mathematical and the physical halves of *Paradoxien des Unendlichen*. While the physical part is not only now out of date, but had little effect on physics (or metaphysics) in its own day, the mathematical part went on through Dedekind, Cantor, Russell and others materially to inform subsequent developments in continuum theory, set theory and transfinite arithmetic. Why it did not have these effects more directly and rapidly than it easily could have is explained in the Appendix below.[2]

BIBLIOGRAPHY

WORKS BY BOLZANO

AA Athanasia oder Gründe für die Unsterblichkeit der Seele, Sulzbach: J. E. v. Seidel, 1827 [published anonymously]; 2nd improved and

---

[2] I am indebted to Edgar Morscher for help in sourcing primary literature and to two referees for helpful suggestions for improvement.

enlarged edition (no longer anonymous): Sulzbach: J. E. v. Seidel, 1838; reprint: Frankfurt/M.: Minerva, 1970.
AP Aphorismen zur Physik. In *BGA* Series II: Nachlaß. A. Nachgelassene Schriften. Vol. 12, 3: *Vermischte philosophische und physikalische Schriften 1832–1848*. Edited by Jan Berg and Jaromír Loužil. Dritter Teil, 106–148, Stuttgart-Bad Cannstatt: Frommann–Holzboog, 1978.
BGA *Bernard-Bolzano-Gesamtausgabe.* Stuttgart-Bad Cannstatt: Frommann–Holzboog, 1975 ff.
MWBB *the Mathematical Works of Bernard Bolzano.* Edited and translated by Steve Russ. Oxford: Oxford University Press, 2004.
PdU *Paradoxien des Unendlichen.* Edited by Franz Přihonský. Leipzig: Reclam, 1851; reprints: 1889, 1920, 1955, 1964 and 1975. English translations: *Paradoxes of the Infinite*. Edited by Donald A. Steele. London: Routledge and Kegan Paul, and New Haven: Yale University Press, 1950; and in *MWBB*, 591–678.
RW *Lehrbuch der Religionswissenschaft, ein Abdruck der Vorlesungshefte eines ehemaligen Religionslehrers an einer katholischen Universität, von einigen seiner Schüler gesammelt und herausgegeben*, 3 parts in 4 volumes, Sulzbach: J. E. v. Seidel, 1834 [published anonymously]. Critical edition: *BGA* Series I, Volumes 6–8.
WL *Wissenschaftslehre. Versuch einer ausführlichen und grösstentheils neuen Darstellung der Logik mit steter Rücksicht auf deren bisherige Bearbeiter*, 4 volumes, Sulzbach: J. E. v. Seidel; 2nd improved edition: Leipzig: Meiner, 1929, 1929, 1930, and 1931; reprints: Aalen: Scientia, 1970 and 1981. Critical edition: *BGA* Series I, Vols. 11–14. English translation: *Theory of Science*, translated by Paul Rusnock and Rolf George, 4 volumes, Oxford: Oxford University Press, 2014. References are to the work's sections.

OTHER WORKS CITED

Ganthaler, Heinrich, and Peter Simons. 'Bernard Bolzanos kosmologischer Gottesbeweis'. *Philosophia Naturalis* 24 (1987): 469–75.
Morscher, Edgar. *Das logische An-sich bei Bernard Bolzano*. Salzburg: Pustet, 1973.
Morscher, Edgar. "Robert Zimmermann – der Vermittler von Bolzanos Gedankengut? Zerstörung einer Legende". In *Bolzano und die österreichische Geistesgeschichte*, edited by Heinrich Ganthaler and Otto Neumaier, 145–236. St. Augustin: Academia, 1997.
Přihonský, Franz. *Atomlehre des sel. Bolzano*. Budissin: E. M. Monse, 1857. Reprinted in Přihonský: *Neuer Anti-Kant und Atomlehre des seligen Bolzano*, edited by E. Morscher and C. Thiel. St. Augustin: Academia, 2003.
Winter, Eduard. "Bolzano in Těchobuz-'Friedenstal'. Ein vormärzliches Idyll unter Polizeiaufsicht". In *Bernard Bolzano 1781–1848. Studien und Quellen*. [No editor given], 279–336. Berlin: Akademie-Verlag, 1981.
Zimmermann, Robert. *Leibnitz und Herbart. Eine Vergleichung ihrer Monadologien. Eine von der königl. dänischen Gesellschaft der*

Wissenschaften zu Kopenhagen am 1. Jänner 1848 gekrönte Preisschrift. Vienna: Braumüller, 1849.

Zimmermann, Robert. *Philosophische Propadeutik für Obergymnasien. Zweite Abtheilung: Formale Logik für Obergymnasien.* Vienna: Braumüller, 1853.

Appendix. Bolzano and Herbart

Since Herbart's monadological views are considered elsewhere in this collection, it is worth saying a little about Bolzano's relationship to Herbart and his views. Directly, there is relatively little to report. Herbart being one of the most prominent German philosophers of his day, the conscientious reader Bolzano was well aware of Herbart's views. Herbart is mentioned a number of times in the *Wissenschaftslehre*, mostly in connection with the theory of ideas and judgements, and mostly critically. Despite ranking Herbart with Fichte, Schelling and Hegel as one of the philosophers he finds make 'utterances so extraordinary to me that I doubt whether I even understand their correct meaning' (*WL* 8), Bolzano nevertheless accords Herbart the epithets 'acute' (*WL* 52) and 'estimable' (*WL* 64), and this, we can trust, without irony. In supporting the idea of his student Fesl that there should be a prize competition to find the best positive criticisms of *RW* and *WL*, Bolzano envisaged Herbart as possible chair of the jury, an idea which was undermined by the latter's death in 1841, and later shelved altogether (Winter, 'Bolzano in Těchobuz-"Friedenstal"', 294).

Despite some convergences in their views on the physically real, the only serious comment Bolzano makes about Herbart's metaphysics is to dispute, without much attempt at refutatory argument, since clearly he saw it as too obvious, the latter's contention that a simple substance cannot have more than one quality (AP 8).

Herbart did however have, indirectly and unbeknown to either (since both were by then dead), an extremely negative impact on the reception of Bolzano's thought. Bolzano had hoped to find a worthy disciple to carry forward or at least publicize his work in logic and mathematics, and thought to have found one in the young Robert Zimmermann (1824–98), son of a close friend. Bolzano instructed the 'dear lad' in mathematics, and when the Austrian school reforms of 1849 under Count Leo Thun and Holenstein (who had studied with Bolzano in Prague) were implemented, Zimmermann wrote a school textbook of logic and psychology that incorporated many of Bolzano's ideas (Zimmermann, *Philosophische Propadeutik für Obergymnasien*), though without mentioning him as their author (as indeed Bolzano had cautiously counselled). The book was in use only from 1853 to 1860, when it was replaced by a second edition in which Zimmermann largely erased any traces of Bolzano's work. The motivation for this change of mind was that Zimmermann had become a convert to Herbart's views, having published in 1849 a prize-winning study comparing Leibniz's and Herbart's monadologies (Zimmermann, *Leibnitz und Herbart*), after translating the former's *Monadologie* into German for the first time (1847). As if this were not enough, Zimmermann voluntarily took on the care of Bolzano's vitally important, ground-breaking and in parts almost print-ready mathematical *Nachlass*, which he then proceeded to ignore and sit on for the rest of his own (highly successful) career. Far from promoting Bolzano's work, Zimmermann, under the influence of Herbart, rendered Bolzano and posterity a notorious disservice, delaying by many decades the appearance of many of Bolzano's revolutionary ideas (vide Morscher, 'Robert Zimmermann – der Vermittler von Bolzanos').

ARTICLE

FROM HABIT TO MONADS: FÉLIX RAVAISSON'S THEORY OF SUBSTANCE[1]

Jeremy Dunham

In this article, I argue that in his 1838 *De l'habitude*, Félix Ravaisson uses the analysis of habit to defend a Leibnizian monadism. Recent commentators have failed to appreciate this because they read Ravaisson as a typically post-Kantian philosopher, and underemphasize the distinct context in which he developed his work. I explore three key claims made by interpreters who argue that Ravaisson should be read as a Schellingian, and show [i] that these claims are incompatible with the text of *De l'habitude* and [ii] how they have obscured from view the monadism at the heart of this work. This article is divided into two sections. First, I explain the importance of Victor Cousin and Maine de Biran for the development of nineteenth-century French philosophy. Second, I argue that to understand the structure of *De l'habitude*, it should be read as a critique of Cousin's philosophical method and a demonstration of the superiority of Biran's Leibniz-inspired introspective method. Like Biran, Ravaisson believes that the introspective method leads to a pluralist metaphysics of forces, but he uses the introspective analysis of habit to go further back to Leibniz than Biran does and develops a pluralist substance metaphysics.

Nineteenth-century philosophy, especially nineteenth-century European philosophy, is commonly referred to as 'post-Kantian'. This label emphasizes the revolutionary impact Kant's critical work had on the practice of philosophy and that the problems most frequently engaged with during

[1] I am very grateful to audiences in Aberdeen, Dundee, and Edinburgh for feedback on earlier versions of this paper. In particular, I must thank Mogens Laerke, Paul Lodge, Beth Lord, and Pauline Phemister. I am also very grateful to Iain Hamilton Grant and Andrew Pyle for their comments, advice, and encouragement on earlier drafts of this article. Finally, I would like to thank the two anonymous referees for their very insightful and encouraging comments, many of which will greatly influence my future research projects. This research was supported by a Postdoctoral Fellowship award from the Institute for Advanced Studies in the Humanities, University of Edinburgh, and an Early Career Fellowship awarded by the Leverhulme Trust and the University of Sheffield.

this time were those left by Kant's *Critiques*, much in the same way as the problems dealt with in the early modern period were shaped by Descartes. One of the key assumptions of this story of philosophy's historical progress is that rationalist metaphysics, of the kind that Leibniz's monadology represents *par excellence*, was no longer legitimately pursuable. Vincenzo de Risi sums up this view neatly:

> After [Kant], in fact, no thinker seems to me to have effectively discussed Leibniz's metaphysics as a still active inspiration in cultural terms ... All the following revivals of Leibniz's thought, from formal logic to Husserl's monads, and beyond, have been local interpretations or antiquarian suggestions. For all of them, monadology irretrievably belongs to the past, and is no longer an enemy to fight, nor an ideal to pursue.
> (*Geometry and Monadology*, xvii)

Risi does not argue for this view because there is no need; it seems so obviously true. It is a well-entrenched historical belief. However, the aim of this article is to show that the practice of philosophy in France during the nineteenth century problematizes this neat view for a number of interesting historical and philosophical reasons. Kant's critical philosophy was not received favourably by France's philosophical community until near the very end of the century and the French philosopher who did the most to promote Kant's views, Charles Renouvier (1815–1903), was not only marginalized, but also believed that the critical philosophy could only be defended once modified by a form of Leibnizianism. Accordingly, he referred to his version of the critical philosophy as a *Nouvelle monadologie*. However, in this article, for three reasons, I focus on the case of Félix Ravaisson (1813–1900). First, Ravaisson defends a classical (non-critical) form of monadological metaphysics. Second, recent attempts to interpret Ravaisson's metaphysics have failed to recognize this because they have been misled into thinking he is a 'typically post-Kantian thinker' (Carlisle, *On Habit*, 61). And, third, although few in number, Ravaisson's works had an enormous influence over the development of French philosophical thought. As Parodi (*La Philosophie contemporaine en France*, 29) states, Ravaisson's major works exerted 'an incontestable authority' and became the 'breviaries of every young philosopher'. Through Ravaisson's works, therefore, Leibnizian ideas became deeply engrained into the philosophical culture. The clearest example of this comes from his 1867 *La Philosophie en France au XIXe siècle*. The book itself was a government-funded report commissioned to celebrate the end of Napoleon III's ten-year suspension of the *agrégation*. However, the final chapter is a kind of manifesto, a plea for a 'spiritualist positivism' whose flourishing would be best ensured by following the principles of a form of Leibnizian monadism. Leibniz's monadism, Ravaisson argued, was an *ideal to pursue*. It is difficult to overestimate the importance of this chapter since the Leibnizian ideas are so explicit, and it became essential reading for all philosophy students taking the *agrégation*. Bergson explains this well:

> No analysis can give an idea of these admirable pages. Twenty generations have learned them by heart. They have counted for a great deal in the influence exercised by the *Rapport* on philosophy as studied in the universities, an influence whose precise limits cannot be determined, nor whose depth be plumbed, nor whose nature be exactly described, any more than one can convey the inexpressible colouring which a great enthusiasm of early youth sometimes diffuses over the whole life of man ... The *Rapport* of 1867 gave rise to a change of orientation in philosophy in the university.
> ('The Life and Works of Ravaisson', 284, 290)

In this article, however, I will not discuss the 1867 report,[2] but focus instead on his similarly influential 1838 work *De l'habitude*.[3] Although it was written almost thirty years before the report, I argue that despite the trend in the recent secondary literature to overemphasize the influence of Schelling and to present a misleadingly *naturalist* reading of the text, a correct understanding of *De l'habitude* must recognize that a form of Leibnizian monadism underlies it also.[4]

In order to develop my argument, in the first section of this article I discuss the context of *De l'habitude*. I show that the circumstances surrounding the writing of this work led him to conceal the extent to which it is a powerful critique of the Scottish Common Sense philosophers' philosophical method as adopted also by Victor Cousin (1792–1867). At the time of writing, Cousin's control over the practice of philosophy in France was such that to make this critique explicit would have been professional suicide. This, I argue, is one of the key reasons why this work is often misread. Although *De l'habitude* is enigmatic, I show that once read alongside his *Philosophie contemporaine*, published two years later, the structure of the argument is greatly clarified. As this latter text was written once Ravaisson had given up on a career in academic philosophy, he used the opportunity of its publication to make both the negative and positive arguments in *De l'habitude* clear. The positive argument, I argue, is a plea for the return to the introspective method developed by Maine de Biran (1766–1824); a method that both Biran and Ravaisson believed was inaugurated by Leibniz. In the second section, via a comparative textual analysis of *De l'habitude* and *Philosophie*

---

[2] The Leibnizianism in the 1867 report is absolutely clear. However, as this was written almost 30 years after *De l'habitude* I will not refer to it in my reading.

[3] Carlisle and Sinclair's excellent recent translation of and commentary on *De l'habitude* and the series of articles they have individually written following its publication have been responsible for a small Ravaisson renaissance.

[4] I doubt anyone will find my reading more appealing as a contemporary philosophical position, but it is, I believe, historically more accurate. Crucially, it allows us to better understand the importance of the contributions of later French philosophers, such as Émile Boutroux and Léon Dumont, who do develop Ravaisson's theory of habit in a more naturalist direction.

contemporaine, I present three key claims made in the Schelling-influenced interpretations of the former work and argue that all of them are inconsistent with the text itself. In contrast, I show that Ravaisson's defence of Biran's introspective method leads him to defend, like Biran himself, a Leibnizian pluralistic metaphysics of forces. Unlike Biran, however, he believes this method also shows us that a true metaphysics must resurrect not just a theory of forces, but a Leibnizian theory of substances too. One of the astonishing results of the introspective analysis of habit is, therefore, that it leads us to a monadological metaphysics.

1. LEIBNIZIAN SPIRITUALISMS: VICTOR COUSIN AND MAINE DE BIRAN

In 1839 Ravaisson wrote to his friend Edgar Quinet that he intended to 'plough the abandoned field of metaphysics' ('Lettres de Ravaisson, Quinet et Schelling', 500). When first considered, the claim that metaphysics had at this time been abandoned by French philosophy seems questionable. From 1830 the eclecticism of Victor Cousin – the philosopher of the restoration – had come to dominate the French philosophical scene and his philosophy had a metaphysical Absolute at its heart. So what did Ravaisson mean? Ravaisson's implicit claim is that Cousin's metaphysics is not worthy of the name. He believed that Cousin unduly restricted himself to following the philosophical method of the Scottish common sense school. Their method, Ravaisson insists, is the Baconian method of induction and observation, which is entirely unsuitable for enquiry into being *qua* being. In order to plough the abandoned field of metaphysics, Ravaisson believed it was necessary to turn away from this method and follow instead the introspective method of his recent French ancestor Maine de Biran. I want to show that understanding this point is crucial for properly understanding the structure of *De l'habitude*'s argument. To make this clear, in this section I introduce some of Cousin's and Biran's key ideas and explain their importance for French philosophy at this time.[5] This will form the essential context for the interpretive argument I develop in Section 2.

Although Cousin did not leave behind a particularly impressive philosophical system of his own, his importance for the development of French philosophy during the nineteenth century should not be underestimated. Between 1830 and 1848 he gained almost complete control over the direction of France's philosophical education. He had, as Jules Simon reports first-hand, a 'despotic authority' of a kind that Hegel, Leibniz, or Descartes could never have dreamed (*Victor Cousin*, 142). Cousin referred to the

---

[5] I have written on the relationship between Cousin and Biran in further depth elsewhere. See Dunham, 'Universal and Absolute Spiritualism'.

instructors of philosophy as his 'regiment' and they formed a regiment over which he had 'every hold':

> He knew the name and the record of each of his soldiers ... If one of these teachers published a review, an edition, an article of any moment, and especially if he published a book, Cousin at once read it ... If the performance was worthless, the man was lost; if there was any trace of talent in it, Cousin became at once his tyrant and his protector. From that time such a person knew no rest until he had shown all that was in him, and had, in return, been provided with a position worthy of his talent. In one way or another there was not a teacher ... of philosophy – whom Cousin did not know by heart.
>
> (Simon, *Victor Cousin*, 116)

It is clear then that because of this influence his verdict regarding Kant's philosophy would be of great importance for its reception and his verdict was not good.[6] He claimed that Kant's restriction of 'universal principles' to 'impressions of sensibility' made objective knowledge solely dependent on the individual subject and when combined with the unknowability of external objects led to the worst excesses of scepticism. Although Cousin marketed himself as having exceptional insight into Kant's thought and wrote books on it, he clearly did not. To some degree this is unsurprising, Kant's philosophy is extremely difficult and Cousin primarily studied a poor Latin translation. Nevertheless, this does show that it is important to keep in mind that we should be careful not to assume that all philosophers in Europe were dealing with the same problems from the same orthodox background.

Cousin believed that the consequences of empiricist, materialist, and critical philosophies of the eighteenth century had been disastrous. The 'age of criticism and destructions' had 'let loose tempests'. The aim of the nineteenth, he claimed, should be 'intelligent restorations' (*Lectures on the True, the Beautiful, and the Good*, 31). Such restorations would bring together the ideals of the French revolution, namely freedom and equality, with principles essential for the stability of the nation, such as a belief in immutable principles of truth, beauty, and goodness. To defend such principles, and their existence as universal and necessary, we cannot settle at the individual, but we must argue for the existence of a ground that is as universal and necessary as they are that acts as the source of both these

---

[6]Kant's philosophy had up to this point received a mixed reception in France. While there were philosophers who took great interest in his work (such as Joseph Marie Degérando), few had any real grasp of its full importance. In Madame de Staël's pioneering and influential work *De l'Allemagne*, she wrote that '[n]o one in France would give himself the trouble of studying works so thickly set with difficulties as those of Kant' (MDS III 96). Certainly, Biran for the most part ignored it and the only work of Kant he read was the 1770 inaugural dissertation.

principles and of finite being themselves. This is the ground he refers to as the 'Absolute'. Although the use of this term is likely to lead the reader to think of the metaphysical systems of Schelling and Hegel, the philosopher whom Cousin draws on most explicitly and quotes from is in fact Leibniz (see *Lectures on the True, the Beautiful, and the Good*, 98–9 and Leibniz's NE 157, 158, and T§189). The Absolute is, he claims, the theory of the realm of ideas, introduced by Plato, but crowned and completed by Leibniz's theory of God.

Cousin referred to his philosophy as 'spiritualist' but his was not the only form of spiritualism on the market. Biran had also developed a form of spiritualist philosophy, but, unfortunately, few of Biran's writings were published during his lifetime and, therefore, the importance of his alternative was only known by a select few. This was in no small part due to Cousin. He inherited Biran's writings after his death and despite the riches reputed to be contained within these manuscripts, delayed publication for ten years. Even then he published at first only one volume; a volume he claimed (falsely) to contain Biran's thought in its entirety (FP III 63). Vermeren ('Les aventures de la force active en France', 159) offers two reasons for this delay. First, Cousin wanted to retain the glory of being the philosopher who first overturned eighteenth-century sensualism; and, second, he feared that Biran's spiritualism would contest the hegemony of his own. As I show below, Cousin was right to be afraid. It was exactly by appropriating Biran's method of philosophizing that Ravaisson successfully overthrew Cousin's philosophy.

The writings of Biran available to Ravaisson provide only a partial picture of his thought, and they come from very distinct periods of his philosophical development. In fact, part of Ravaisson's originality comes from his ahistorical treatment of Biran's work and the way he consequently brought the insights from the distinct periods together into a systematic metaphysics of forces. The two texts that I believe were especially important for Ravaisson – *Influence de l'habitude sur la faculté de penser* (1799) and *Exposition de la doctrine philosophique de Leibniz*[7] (1819) – exemplify this clearly. In the first, despite its impressive originality, Biran's empiricist and ideological lineage is clear and he maintains that we can know 'nothing of the nature of *forces*' (*The Influence of Habit on the Faculty of Thinking*, 52). Nevertheless, he still insists that reflection on the principle of habit reveals an active psychological principle of voluntary attention. For Biran, there is always a mutual correspondence between the active and the passive in our impressions. Yet when activity dominates, we *perceive*, while when passivity dominates, we merely *sense*. The law of habit is that with repetition our perceptions become clearer and subtler, but our sensations, on the contrary,

---

[7] In *De l'habitude* itself Ravaisson references *Influence de l'habitude* and Biran *passim*. However, we know that Ravaisson knew the *Exposition* well and that it was important for the development of his thought from his 1834 *Mémoire* on Aristotle.

degenerate: 'the less we *feel*, the more we *perceive*' (*The Influence of Habit on the Faculty of Thinking*, 87). From this 'double-law' of habit, Biran concludes that there must be two distinct perceptual faculties, one active and the other passive, since if we attempted to reduce our explanation to one alone, there should be no reason for why certain impressions degenerate and others increase in clarity. In later works Biran will argue that the active psychological principle is a 'hyper-organic' force distinct from physiological workings (see OMB VII 169–70). As we shall see below, this 'double law' of habit and the conclusion that its operation requires a *hyper-organic* cause is crucial for Ravaisson (see RH 37 and 49).

By the time of the (1819) *Exposition* there had been a great transformation in Biran's interests and he gained a confidence to engage in metaphysical speculation that he did not have twenty years earlier. He postulates the existence of a 'universal and absolute spiritualism' and a 'science of forces'. The shift in Biran's views came from his lengthy engagement with Leibniz's philosophy. Despite the fact that Leibniz is sometimes regarded as one of the clearest examples of a 'rationalist' philosopher, it is in his works that Biran believed we could find the key to the reformation and improvement of empiricism. The more he studied Leibniz's philosophy, the more he believed could be gained from the empiricist point of view if experience is understood in its broadest possible sense.

Biran believed that Descartes should be regarded as the true father of metaphysics, since he made the testimony of inner sense the generative principle of all knowledge (OMB VI 17–18). Even so, he argued that Cartesianism is fundamentally flawed insofar as it ultimately leads to pantheism. This is because created Cartesian substances are merely passive. While Descartes is explicit about this in the case of extended substance – the only qualities that belong to its essence are extension, flexibility, and changeability (AT VII 31: CSM II 20) – it follows from his claim that the distinction between creation and preservation is only 'conceptual' that, ultimately, thinking substance must be merely passive too. If the same force initially needed to create the world (the infinite force of God) is required at every moment to preserve it (AT VII 48–9: CSM II 33), then whenever it feels as though I voluntarily will an action, it is not the I that wills, but God. I have the desire (itself caused by God), but I am not responsible for the causal action. Biran claims that it is 'logically certain that all effects are eminently or formally enclosed in their cause' (OMB XI-I 142), so if God is the sole cause, it follows that every created being is enclosed in God, and there is no real distinction between God and nature. Leibniz's great merit, Biran insists, was to be the only early modern philosopher able to present a true metaphysics of personality in which individuality and freedom need not be undermined by the power of God:

> To what did Leibniz grasp onto to keep himself from this dangerous precipice, which, since the origin of philosophy, has led the boldest and most profound

speculators towards the empty concept of the *great whole, nothingness* deified, the devouring abyss that comes to absorb all individual existence? We must say it, the author of the system of monads was saved from this disastrous aberration only by the nature or the proper character of the principle on which he based his system; a principle truly one and individual – the primitive fact of the existence of the *I*, before having acquired a unique and absolute notion. A system that multiplied or divided the living forces in accordance with the intelligible elements or atoms of nature, would, it seems, prevent or dissipate forever those sad and disastrous illusions of Spinozism, too favoured by Descartes's principle.

(OMB XI-I 140)

Although, on the final analysis, all of Descartes's created substances were ultimately passive, for Leibniz they are all active. Biran saw Leibniz's philosophy as providing a true metaphysics of forces. For Leibniz, rather than substance being a placeholder for forces in which they inhere, force *constitutes* substance. While Descartes 'constructed thought with elements borrowed from a passive nature', Leibniz 'constructed nature with elements taken from the activity of the I' (OMDB VIII 223). Biran places a great deal of importance on a 1694 text called 'On the Corrections of Metaphysics and the Concept of Substance' (G IV 468–70: L 432–4). Leibniz there claims that active force

> contains a certain act or entelechy and is thus midway between the faculty of acting and the act itself and involves a conatus. It is thus carried into action by itself and needs no help but only the removal of an impediment.

This text is so important because in it, as well as in texts like *On Nature Itself* (see G IV 510: AG 161), Leibniz appears to establish his conception of active force from introspection. We understand the nature of force through a posteriori reflection on our self-activity. Furthermore, he then applies the principle of uniformity. This states, when Leibniz makes it explicit, that 'all the time and everywhere everything's the same as here' (G III 343: WFNS 220–1). By reflection on our own activity and then by analogy, therefore, we are not only able to discover truths about the nature of created substances, but even about the ultimate substance: God.[8] Biran does of course recognize that there is a clear rationalist side to Leibniz's work, but he argues that this side is in fact *incompatible* with the empiricist side. When Leibniz argues from introspection we arrive at a dynamic metaphysics of forces capable of providing a real ground for freedom and individuality, but when Leibniz argues from

---

[8]Biran is right to highlight these often-neglected a posteriori arguments found in Leibniz's metaphysics. For further details of these arguments in Leibniz's philosophy, see Lodge, 'Leibniz on Created Substance and Occasionalism' and Phemister, 'All the Time and Everywhere Everything's the Same as Here'. I address Biran's use of Leibniz, how the infusion of Leibniz into French philosophy came about, and the particular distinctive character it developed from this infusion in more depth in Dunham, 'Universal and Absolute Spiritualism'.

a priori reason he ends up with an all-encompassing God which subsumes all individuality just as surely as the God of Descartes, Malebranche, and Spinoza. Therefore, the empiricist side provides us with a foundation for the true science of the human mind, and the rationalist side must be rejected. This division of Leibniz's philosophy into rationalist and empiricist sides was hugely important for the French Leibniz renaissance that followed, and the influence of Biran's empiricist Leibnizianism can be seen in, amongst many others, the works of Renouvier, Émile Boutroux, Émile Boirac, and, most importantly for this article, Ravaisson. The two-step process of introspection and analogy is the method Ravaisson will use for his metaphysical argument for monads in *De l'habitude*.

2. MONADISM IN DE L'HABITUDE

In this section, I argue that in *De l'habitude*, Ravaisson uses the Biranian introspective method to argue, via an analysis of habit, for a pluralist metaphysics strongly influenced by Leibniz's metaphysics of monads. However, the philosopher whose metaphysics commentators most frequently align Ravaisson's with is not Leibniz, but Schelling. The thought that we should understand Ravaisson's philosophy in terms of Schelling's influence has been the subject of debate in the French literature,[9] but it has been consistently adhered to in the Anglo-American commentaries on Ravaisson. Although I do not deny that Schelling had some influence on his work, the attempt to interpret Ravaisson as a Schellingian has been a constant source for the misinterpretation of *De l'habitude*. I shall now address three particularly important examples to illustrate this. I focus on these because I believe they have played a significant role in obscuring from view the monadism at the heart of Ravaisson's philosophy. I will refer to them as 'Schelling-Influenced Interpretation' (SII): 1–3:

> SII1: George Boas (1929, 16) suggests that Ravaisson turned Biran's theory of Will into 'a sort of Schellingian Absolute'.
>
> SII2: Carlisle and Sinclair ('Commentary on *Of Habit*', 84) claim that for Ravaisson, '[t]he development of habit, along with that of life and nature *emerges* from the inorganic'.
>
> SII3: Sinclair (Ravaisson and the Force of Habit, 8) argues that '[r]eflection on motor habits allows Ravaissson in a manner that is doubtless influenced by ... Schelling ... to conceive freedom not as opposed to nature, but rather as *inhabiting* or *animating* the natural body'.[10]

[9]See, for example, Dopp, *Félix Ravaisson*; Janicaud, 'Victor Cousin et Ravaisson, Lecteurs de Hegel et Schelling'; Mauve, 'Ravaisson lecteur et interprète de Schelling'; Guibert (2006).
[10]This is on the list of Schelling-influenced interpretations because the commentator has referred to it as such. However, I am not only sceptical with regard to the claim that there is something corresponding to a force inhabiting or animating a natural body in Ravaisson's

All three of these claims rely on an assumption that understanding Schelling, who is not cited once in *De l'habitude* itself, is the key to understanding it. However, I shall show that all three are incompatible with the text itself. It is true that there are many interpretive difficulties involved in reading *De l'habitude*. Although it is a short text (it takes up less than 30 pages in the most recent edition), it deals with a very broad range of philosophical issues. Crucially, it was Ravaisson's doctoral thesis. As we may understand given Cousin's despotic control, it was necessary for him to ensure that any directly critical points concerning the despot's own work were sufficiently hidden. However, I argue that such critical points are certainly to be found within it, but to identify them *De l'habitude* should be read alongside his 1840 *Philosophie contemporaine*. This latter text was published two years later, therefore *after* he had been awarded his doctorate. By this time he had taken up the position of *Inspecteur général des bibliothèques* and no longer had any desire to be a part of Cousin's university system. Although ostensibly *Philosophie contemporaine* is an article on a French translation of some of William's Hamilton's works, Ravaisson uses the opportunity to criticize Cousin's method, to reproach Cousin for failing to understand the importance of Biran's method, and to explain some of his key arguments first expressed, although less explicitly, in *De l'habitude*. One important aspect of *De l'habitude* that becomes clear when read alongside *Philosophie contemporaine* is the distinct aims of its two parts.

2.1. The Objective and Subjective Analyses

De l'habitude is split into two main parts and the second is much longer than the first. Gabriel Madinier (*Conscience et mouvement*) has helpfully referred to these as the 'objective analysis' and the 'subjective analysis', respectively (hereafter OA and SA). According to my reading the aim of the OA is to establish what knowledge can be acquired about the nature of habit from the external senses, that is, through the Baconian experimental method adhered to by the Scottish school of William Hamilton, Thomas Reid, and Dugald Stewart. According to Ravaisson, this school was 'founded on the idea that it is necessary to extend Bacon's theory, which relates to the sciences in general and to physics in particular, to philosophy' ('Philosophie contemporaine', 398).[11] A critique of this position is important for Ravaisson because he argues that Cousin himself 'does not at all exceed the

---

ontology, but also sceptical about the claim that there is something corresponding to this in Schelling's. Schelling even writes that 'Life-force, taken in this sense, ... is a *completely contradictory concept*' ([1803] 1988, 37). He rejects the idea of 'Life-force as a spiritual principle' because 'how a mind can act physically we have not the slightest idea' ([1803] 1988, 38). For a summary of Schelling's *non-Aristotelian* metaphysics of an antithesis of forces, see Dunham, Grant, and Watson (2014, 129–43).

[11] All translations of Ravaisson's 'Philosophie contemporaine' are by Jeremy Dunham and Mark Sinclair and taken from Ravaisson, 'Contemporary Philosophy'.

limitations of the Scottish speculations' ('Philosophie contemporaine', 408). This theory fails because it limits itself to 'generalities' that do not properly constitute *science*. A proper *science* should be concerned with the *how*, the *why*, and not just the *that*. Part of the problem is that the Scottish method fails to distinguish between mere perception and apperception. The superiority of Biran's method comes from his clear understanding of this Leibnizian distinction and his introduction of the analysis of *inner phenomena* as a science of first principles. These are linked because, Ravaisson claims, 'complete knowledge, human knowledge, is not at all simple *perception* applied to the external object, as it is in the animal; it is reflective perception, Leibniz's *apperception: perception with reflection conjoined*' ('Philosophie contemporaine', 419). Apperception is 'experience of a cause', and philosophy, if done properly, should be 'the science *par excellence* of *causes* and the *spirit* of all things, because it is above all the science of inner Mind in its living Causality' ('Philosophie contemporaine', 420). The aim of the SA is to establish how much more can be discovered about habit from this Biranian perspective. It is only from this perspective that we can move from the mere *that* to the *how* and the *why*, and it is Cousin's great fault to have failed to perceive the fecundity of Biran's method. In the language of Leibniz's *New Essays,* although Cousin had claimed to have been the one who had overturned eighteenth-century sensualism, Ravaisson accuses him of having gone no further than Lockean 'nominal definitions' and of having failed to reach Leibnizian 'real, causal' definitions which would allow us to understand the 'structure or inner constitution' of being (NE 294).

The OA begins with an analysis of the inorganic, and then moves through the stages of being through vegetal, animal, and finally human life. The inorganic realm displays no evidence of habits whatsoever, a very limited form of habit is displayed in vegetal life, a superior degree is found in animal life, but it is when we reach human life that we discover habit proper as directed by activity. Evolutionary interpretations such as SII2 rest on the false assumption that Ravaisson is developing a positive theory of habit in this part and from a failure to distinguish the distinct roles of each analysis.[12] However, correctly understood, the aim of this first analysis is to make the inadequacy of the Baconian method clear. Before the OA properly begins, Ravaisson sets out some basic definitions of 'habit' and 'being' that when combined with his discussion of the inorganic at the beginning of the OA show that he does not regard the inorganic as the foundation for the development of the rest of life, but rather as a realm that has in itself no true

---

[12]If we fail to identify the distinct roles of each analysis, then Ravaisson would appear to make contradictory statements. For example, as Bruyeron, 'Remarques sur un passage du texte de Ravaisson' has noted, in the OA Ravaisson writes that habit is not possible in the inorganic realm (RH 29), but in the SA he writes that habit, thought, and activity are traceable even in crystals (RH 67). However, instead of recognizing the distinct roles of each analysis, Bruyeron provides a highly speculative attempt to make the two claims consistent. I cannot see how any such attempt could work.

being, since true being must be grounded in determinate substances. Reconstructed, his argument appears remarkably similar to Leibniz's famous argument for monads that opens the latter's *Monadology*. The suspicion that Ravaisson has this argument in mind is further aroused by Ravaisson's support of the 'Universal Law of Being', which he attributes to Leibniz, and describes as stating that 'the fundamental character of being, is the tendency to persist in its way of being' (RH 27). His argument is as follows:

(1) The primary realm of nature (the inorganic) is one of homogeneous extension.
(2) Extension is divisible to infinity.
(3) Causality, energy, and potential must reside in a *determinate substance*.
(4) Only a true unity can be a determinate substance.
(5) There are no true unities in the inorganic realm if it is understood as homogenous extension [from 1 and 2]
(6) Without causality, energy, and potential, an entity cannot strive to persist in its own being.
(7) The 'Universal Law' states that only those entities that strive to persist in their own being possess the 'fundamental character of being'.
(8) Therefore, the inorganic does not possess the fundamental character of being. [by 3, 4, 5, 6, and 7]

Habits cannot belong to mere extension alone because a necessary condition of habit is change. Ravaisson claims that habit 'supposes a change in the disposition, in the potential, in the internal virtue of that which does not change' (RH 25). As completely homogeneous, extension is incapable of real change. Habit requires an 'individual being' as its ground, that is, a determinate heterogeneous whole that cannot, unlike extension, be divided; in other words, a substance. Therefore, this blocks SII2 because habits must belong to substances, and substances cannot 'emerge' from the inorganic. Habits require time and change, and these must be grounded in determinate substances that do not change. Substances cannot 'emerge' from the inorganic because the emergence of substances from the inorganic in Ravaisson's system would be as absurd as the emergence of monads from their phenomena in Leibniz's. Furthermore, such a view anachronistically attributes a view of the evolution of life that Ravaisson simply does not hold. In fact, in an undated fragment, he wrote that 'one would search in vain to prove by experience that which we call the spontaneous generation' of beings endowed with 'life, feeling, and thought' from inert bodies (Janicaud, *Ravaisson et la métaphysique*, 243).

If the inorganic realm lacks true being, what kind of reality does extension have? Ravaisson is quite clear about this. Extension, like time, is one of the *conditions* under which we *represent* substances (RH 27). We represent substances through the phenomenal forms of space and time as 'extended mobiles', but extension itself has no independent existence outside of the substances themselves. Extension is a 'sensible form' that is necessary so that the understanding may represent quantity to itself (RH 41).[13] As he puts it explicitly in the SA:

> In order to represent the synthesis of diversity in space, it is necessary not only that I be a substantial subject accomplishing movement, at least in the imagination, but also that I conceive it, that I mark its end, and that I will its direction.
> (RH 41)

Now we can see why SII3 is also blocked as an interpretation of Ravaisson's philosophy of nature. Freedom cannot be a force *inhabiting* a natural body, because body is not the sort of thing that can be inhabited or animated. Ravaisson is an idealist about bodies in the way that certain commentators refer to Leibniz as an idealist. For Ravaisson, bodies are well-founded phenomena. We represent non-spatial substances to ourselves via the sensible form of extension. When we represent these substances to ourselves as extended mobiles, these mobiles are *expressions* of the true metaphysical substances. This is the sense to his claim that 'the Mind is not invisible, but the only visible' ('Philosophie contemporaine', 427). Leibniz scholars who reject the idealist interpretation of his metaphysics sometimes turn to his relationship to Aristotelian metaphysics in order to stress the importance of the coexistence of entelechies and bodies (see, for example, Phemister, *Leibniz and the Natural World*, 20–1). Those who may wish to reject my idealist interpretation of Ravaisson might be tempted to use this strategy, since he wrote several volumes on Aristotle's philosophy. However, this approach would not work because Ravaisson reads Aristotle's matter as neither physical nor corporal but rather reducible to 'pure logical possibility'.[14] In his unpublished academy prize winning 1834 memoire[15] on Aristotle's philosophy, he claimed that:

> matter is only a relative term without reality in itself, and by descending by continual degrees *we would reduce it to pure abstract possibility, which is no longer anything more than a moment of being, a point of view under which it* [being] *is considered.*
> (1834, 164, cited in Dopp, *Félix Ravaisson*, 98)

---

[13]Ravaisson's use of Kantian language such as the 'intuition of space' or 'sensible form' has understandably led commentators to interpret him as very close to Kant on these points (see Lovejoy, 1913, 469–70). However, we will not learn much about Ravaisson from such comparisons. He is dismissive of Kant's critical philosophy and claims that '[i]n Kant's system, being is the deceptive image of the empty form that we call time, and it is the dream of the intellect that takes this nothing for a thing' ('Philosophie Contemporaine', 412). As Bellantone, 'Ravaisson: le 'champ abandonné de la métaphysique'' notes, one point on which both Cousin and Ravaisson were agreed on is their rejection of Kant and his 'scepticism'. It is important for Ravaisson that through time and space we represent things in themselves.
[14]See Cousin, *De la Métaphysique d'Aristote*, 96–7 and Dopp, *Félix Ravaisson*, 97–8.
[15]Although the whole text is unpublished, selections have been published in Cousin's 1838 report on the contest (90–119) and in Dopp, *Félix Ravaisson*, 80–124).

Therefore, to understand the true nature of beings, we must turn away from the evidence of the external senses and reflect on our own essence. We must turn to the 'subjective' analysis. At the end of the OA, following the first reference to Biran, Ravaisson tells us that:

> Up to this point, nature is for us a spectacle that we can only see from the outside. We see only the exteriority of the actuality of things; we do not see their dispositions or powers. In consciousness, by contrast, the same being at once acts and sees the act; or better, the act and the apprehension of the act are fused together. The author, the drama, the actor, the spectator are all one. It is, therefore, only here that we can hope to discover the principle of actuality.
>
> (RH 39)

2.2. Biranian Introspection and Substances

The superiority of the SA over the OA is a point that Ravaisson further clarifies in *Philosophie contemporaine* where he tells us that a true understanding of unity, causality, and substance could only be obtained through the Biranian introspective method. 'To remain any longer submitted to the foreign doctrine [the Baconian method]', he says, 'would truly be *inventa fruge, glandibus vesci* [to feed on acorns, when corn has been discovered]'[16] ('Philosophie contemporaine', 423). Crucially, Ravaisson believes that the Biranian method shows that what he calls the 'Scottish Axiom' – that 'experience attains neither causes nor substance' – is false. However, it is important to note that, unlike Ravaisson, Biran would have agreed with the second half of this axiom. Although Biran believed that introspection provided us with evidence of force and causality, he did not believe that the concept of 'substance' had any further meaning beyond these forces. He thought of substance as a 'passive subject of modifications' and Biran rejected the existence of anything passive in his metaphysics. Ravaisson, on the contrary, argued that the rejection of substance would lead to a metaphysics of the soul where beyond the knowledge of our immediate willings, we are left with an impenetrable abyss. When separated from substantial reality, he claims, the will is 'only an abstraction' ('Philosophie contemporaine', 425). The will is not groundless, but is the product of anterior tendencies grounded in a true substantial unity. Ravaisson backs this Leibnizian point up by two quotes from Leibniz that are very telling. The first is from *Of Nature Itself*:

> We must add a soul or a form analogous to a soul, or a first entelechy, that is a certain urge [*nisus*] or a primitive force of acting, which is itself an inherent law [*lex insita*], impressed by divine decree.
>
> (G IV 512: AG 162–3)

---

[16]This is a nod to Leibniz's fifth letter to Samuel Clarke (AG 344).

The second is from a letter to Bayle where Leibniz writes that 'the nature of substance consists, in my opinion, in this ruled tendency from which phenomena arise in order' (G III 58). These two passages are crucial because they clarify exactly what Ravaisson thinks is missing from Biran's account, and this is necessary since Leibniz himself does at times seem to suggest that substance is nothing over and above force; he writes that force is 'what *constitutes substance*, since it is the very principle of action, which is its characteristic feature' (G IV 472: WFNS 22). Nonetheless, Leibniz makes a distinction between primitive and derivative forces, which constitute the atemporal and temporal sides of a substance respectively. Primitive active force, limited by primitive passive force, together forms a complete monad (LDV 261 and 265). If a monad were purely primitive active force, it would be exactly like God. Primitive passive force limits a monad's nature, and is the reason why it is distinct and inferior to God. Together, Leibniz tells us, these forces form the 'complete monad' because they constitute the non-temporal foundation from which the temporal series of perceptions arise. The relationship between the primitive forces in a monad does not change: 'the soul remains the same' (LDV 77). As Leibniz writes 'primitive force is like the law of a series, and derivative force is like a determination that designates some term in the series' (LDV 287). For Ravaisson, Biran eliminates primitive force leaving us with the derivative forces alone, but he leaves these latter ungrounded. As we have already seen, for Ravaisson that which changes requires a ground which does not change, and this is what exactly Leibniz's primitive forces provide. In *De l'habitude* Ravaisson refers to this as 'primitive nature' and argues that it is in this nature that our voluntary nature has its 'source and origin' (RH 61). Ravaisson's defence of this aspect of Leibniz's philosophy shows how far Ravaisson was from the true voluntarist French philosophers of the nineteenth century like Biran and Renouvier, but also how far he was, at least in terms of this fundamental metaphysical belief, from Schelling. In fact, Schelling says that 'the person who cannot think activity or opposition without a substrate cannot philosophize at all' (*First Outline of a System of the Philosophy of Nature*, 219).

2.3. The Analysis of Habit and the Metaphysics of Monads

Although the discussion in Section 2.2 provides sufficient evidence to undermine SII1, if understood in its strictest sense, Boas actually only claims that Ravaisson transforms Biran's will into a *sort of* Schellingian Absolute. Charitably understood then, he may just be claiming that Ravaisson has developed a form of monism. However, in what I think is the most interesting part of the SA, he presents an introspective argument for a substance pluralism. In this final section, I show that this conclusively undermines SII1. The analysis of habit, Ravaisson argues, leads to a pluralist metaphysics of monads. This analysis is original to Ravaisson, but it leads him to a Leibnizian metaphysics.

When introducing the SA and its importance for a full understanding of habit, Ravaisson writes that:

> [i]t is in consciousness alone that we can find the archetype of habit; it is only in consciousness that we can aspire not just to establish its apparent laws but to learn its *how* and its *why*, to illuminate its generation and, finally, to understand it cause.
>
> (RH 39)

This is a dense passage. First, it distinguishes the OA, which established *that* there are laws of habit (especially the 'double law' discussed in Section 1), from the SA that will seek their *how* and *why*. Second, by emphasizing this distinction it points to Schelling's critique of Cousin who wrote of the latter's philosophy that 'it is limited to generalities that do not promise in the least... a science properly speaking' ([1835] 1988, 38). 'It is content with the *that* without struggling for the *how*' ([1835] 1988, 72). Ravaisson certainly was influenced by Schelling's critique of Cousin and maintained in line with it that to settle at the *that* would be 'to banish from philosophy the very object of any philosophy worthy of the name' ('Philosophie contemporaine', 402). However, in the SA we not only learn about the *how* and the *why* of habit, but also that habit itself is 'the most powerful of all analogies' (RH 65). This is because its analysis connects our conscious activities with our most elementary mental activities and even the most elementary modes of existence. Understanding habit leads us to an understanding of our *primitive* nature.

Following Biran, Ravaisson argues that the double law of habit – which states that 'continuity or repetition dulls sensibility, whereas it excites the power of movement' (RH 53) – demands a hyper-organic cause to explain its *how*. Consider, for example, attempting to learn to play a thirteenth-century piece *Como Somos per Consello* on the mandolin and trying to perfect its haunting melody. If I tried to do this, at first the result would be disappointing. It would take a while for my hands to correspond in the correct way to play the tune as well as I would like. As I actively practice the tune, my playing will become more precise. I will be able to notice more distinctly the slight variations in the length of the notes, which help to provide the tune with a haunting feeling. Over time I will become increasingly sensitive to the ways I can express the emotive nature of the tune in a consistent manner. The force of habit in relation to our active and willed movements reinforces these actions and allows us to act in a way that is more prompt and precise. At a certain point, I will be able to play the song without concentrating on it. In fact, I will play the song better if I do not. Such an ability will be necessary to acquire if I wish to be able to sing a melody at the same time as playing the mandolin. At this point, effort has faded and habit has taken over. For Ravaisson, this 'acquired *tendency,* the inclination whose progress coincides with the degradation of sensation and effort', is exactly what cannot be explained by any 'organic modification' (RH 53). As he goes on to explain, this is because even when I play the tune without consciously focusing on it, there is still an intelligent process going on. The distinction between 'goal' and 'action' has been effaced, but intelligence has not been eliminated, it is still the necessary condition for the behaviour to be manifested because it is an intentional action and it acts in accordance with an idea, even if this idea is being perceived below the level of consciousness.

This process is similarly reflected and Ravaisson's argument can be made clearer in cases of 'bad habits' like a compulsive cough or clearing of the throat. At first this may begin because of an irritation of the throat, but frequent repetition may form a habit that becomes an agonizing psychological nightmare to attempt to resist; it will feel as if one had no choice but to do it. At a certain point this behaviour may become almost completely concealed from the subject, to such a point that they would be shocked to hear an audio recording of themselves. In such behaviour what had begun as voluntary behaviour guided by the will has been transformed by habit into inclinations, perhaps even instincts, which precede the will and consciousness. For Ravaisson, '[c]onsciousness implies knowing, and knowing implies intelligence' (RH 39), but intelligence implies neither knowing nor consciousness, and this means that intelligent processes may occur without conscious awareness. However, it is still goal-orientated behaviour and he claims that even if it does not occur consciously, 'every inclination towards a goal implies intelligence' (RH 53). Since he argues that there is only a difference in degree between instinct, inclinations, and our goal-directed conscious behaviour, even in our primitive instincts there is a kind of obscure perception of an idea that directs their activity and determines them as tendencies.

In the SA and in *Philosophie contemporaine*, Ravaisson is concerned to show that this development of tendencies is dependent on our non-organic substantial being. To explain how Ravaisson conceives this, let us analyse the mandolin example further. To begin to learn how to play the song, we must begin with an idea of it as a worthy end to attain. However, before we have developed the desirable feeling towards this idea and it has become a 'motive', it must, he insists, have previously existed as

> some kind of unreflective and indistinct idea, which occasions reflection and constitutes its matter, its beginning and its basis ... It is in the uninterrupted current of involuntary spontaneity, flowing noiselessly in the depths of the soul, that the will draws limits and determines forms.
>
> (RH 73–5)

Ravaisson calls these unreflective and indistinct ideas *mobiles*, since they have an active influence on our behaviour, but he claims that they are mobiles that 'do not differ from the soul itself'. Before we act through *thought*, they act by *being* ('Philosophie contemporaine', 425). We know from Ravaisson's 'ultimate law' that the very character of being is the tendency to persist in its own being, so this primitive striving is a primitive *force*. Therefore, our decision to choose to learn to play *Como Somos per Consello* requires the *actualization* of an idea that previously had an active and effectual *virtual* existence in our *primitive* nature. However, the full actualization of this idea requires that we 'make it flesh' through practice and effort. At first my idea of myself playing this tune is a mere virtuality, then I consciously represent the idea of playing it to myself, and once I have decided to learn how to play it, and I practice and slowly become better at playing it, I approach this idea more closely. As Ravaisson writes, 'the end becomes fused with the movement, and the movement with the tendency', when this occurs, 'the ideal is realised in it' (RH 55). Through habit, the idea will have become 'more and more a *substantial idea*' (ibid.). What Ravaisson means by this is that the gap between the 'idea' of playing the tune and my effort will have become effaced, our second nature will have fully realized our primitive nature: 'the final degree of habit meets *nature* itself' (RH 61). This is the actualization of what Ravaisson calls 'prevenient Grace'. Our 'primitive nature', he says, is 'God within us' (RH 71); it is the extent to which our own limited essence reflects God's. Here Ravaisson is expressing what Nadler (*Spinoza's Ethics*, 118) calls an *immanentist pantheism*. He believes that God is 'ontologically distinct' from the rest of the substances which make up his universe, yet at the same time God is 'ubiquitously contained' within all of them. Ravaisson's metaphysics is in complete agreement with Leibniz's statement in §47 of the *Monadology* that all created substances 'originate, so to speak, through continual fulgurations of the divinity from moment to moment, limited by the receptivity of the created being, to which it is essential to be limited' (*cf.* T §382–91, 395, 398). God is the primitive unity from which the created substances derive their power. Insofar as they are active, they reflect God's essence. Insofar as they are passive, they are distinguished from God's perfection.

Ravaisson's argument is that reflection on habit leads us to postulate the existence of a unique centre that unites the set of innate ideas which exist as part of our atemporal essence and that are necessary for the formation of our temporal ideas and permanent habits. As there is only a difference in degree between conscious reflection, inclinations, and instincts, habit leads us from our conscious experience to our very primitive nature. However, habit is the 'most powerful of all analogies' because following the use of the Leibnizian principle of uniformity crucial to Biran's method, Ravaisson argues that we can analogously understand the essence of all created beings to share these essential characteristics. Reflection on our organic nature leads us to recognize that in addition to our 'cerebral unity', life diffuses into a 'plethora of independent centres'. These centres are able to develop their own habits and alter their instincts. For example, our muscles become stronger and more agile. Such behaviour would be impossible, he believes, unless there was in these centres a form of obscure perception of ideas. Furthermore, the analogy of habit allows us to descend through the chain of life and form the same conclusions even in the 'deepest heart of nature' and in 'abnormal' and 'parasitic' life (RH 63):

> The most elementary mode of existence, with the most perfect organization, is like the final moment of habit, realized and substantiated in space under a sensible form. The analogy of habit penetrates its secret and delivers its sense over to us. All the way down to the confused and multiple life of the zoophyte, down to plants, even down to crystals, it is thus possible to trace, in this light, the last rays of thought and activity as they are dispersed and dissolved without yet being extinguished, far from any possible reflection, in the vague desires of the most obscure instincts.
>
> (RH 67, *translation modified*)

In conclusion, if habit is a phenomenon which is found in every part of living nature and is in fact contiguous with being, and habit implies obscure perception of an idea, it follows that ideas are not innate in the minds of human beings alone but innate in the substantive essence of every being. The argument from habit in Ravaisson's *De l'habitude* is an argument for a deep *innateness* throughout nature. Every being is in possession of its own set of innate ideas which it strives towards and attempts to actualize according to the rule of the best, that is, it aims to actualize as much of the divine as possible given its limitations. Every being is, for Ravaisson, a monad.

BIBLIOGRAPHY

ABBREVIATIONS
AG: Leibniz, G. W. (1989) *Leibniz: Philosophical Essays*. Translated by R. Ariew and D. Garber. Indianapolis: Hackett.
AT: Descartes, R. (1974–89) *Oeuvres de Descartes*. Edited by C. Adam and P. Tannery. Paris: J. Vrin.
CSM: Descartes, R. (1984–91) *The Philosophical Writings of Descartes*. 3 Vols. Translated by J. Cottingham, D. Stoothoff, and D. Murdoch. Cambridge: Cambridge University Press.
FP: Cousin, V. (1838, tome I; 1840, tome III) *Fragments philosophiques*. Paris: Ladrange.
G: Leibniz, G. W. (1875–90) *Die Philosophischen Scrhiften von Leibniz*. 7 vols. Edited by C. I. Gerhardt. Berlin: Weidmann.

L: Leibniz, G. W. (1989) *Gottfried Wilhelm Leibniz: Philosophical Papers and Letters*. Translated by L. Loemker. London: Kluwer.

LDV: Leibniz, G. W. (2014) *The Leibniz-De Volder Correspondence*. Translated by P. Lodge. New Haven, CT: Yale University Press.

MDS: Staël-Holstein, A. L. G. ([1810] 1814) *Germany*. Translator Unknown. London: John Murray.

OMB: Maine de Biran (1984–2001) *Œuvres de Maine de Biran*. 20 Vols. Edited by F. Azouvi et al. Paris: Vrin.

OMDB: Maine de Biran *Œuvres de Maine de Biran*. Edited by P. Tisserand. Paris: Félix Alcan.

RH: Ravaisson, F. ([1838] 2008) *Of Habit*. Translated by C. Carlisle and M. Sinclair. London: Continuum.

T: Leibniz, G. W. (1985) *The Theodicy*. Translated by E. M. Huggard. Chicago: Open Court.

WFNS: Woolhouse, R. S., and Francks, R. (1997) *Leibniz's 'New System'*. Oxford: Oxford University Press.

OTHER TEXTS CITED

Bellantone, A. 'Ravaisson: le 'champ abandonné de la métaphysique''. *Cahiers philosophiques* 129, no. 2 (2012): 5–21.

Bergson, H. 'The Life and Works of Ravaisson'. In *The Creative Mind*, edited by M. L. Andison, 261–300. New York: Philosophical Library, 1946.

Biran, M. de. *The Influence of Habit on the Faculty of Thinking*. Translated by M. Donaldson Boehm. Westport, CT: Greenwood Press, 1970.

Bruyeron, R. 'Remarques sur un passage du texte de Ravaisson'. In *Ravaisson*, edited by J. M. Le Lannou, 33–46. Paris: Éditions Kimé, 1999.

Carlisle. *On Habit*. London: Routledge, 2014.

Carlisle, C., and M. Sinclair. 'Commentary on *Of Habit*'. In RH, 78–114, 2008.

Cousin, V. *De la Métaphysique d'Aristote: Rapport de le concours ouvert par l'Académie des sciences morales et politiques*. Paris: Ladrange, 1838.

Cousin, V. *Lectures on the True, the Beautiful, and the Good*. New York: D. Appleton, 1854.

Dopp, J. *Félix Ravaisson: La Formation de sa Pensée d'après des documents inédites*. Louvain: Editions de L'Institute Supérieur de Philosophie, 1933.

Dunham, J. 'A Universal and Absolute Spiritualism: Maine de Biran's Leibniz'. In *Maine de Biran: The Relationship Between the Physical and Moral in Man*, edited by D. Meacham and J. Spadola. London: Bloomsbury, forthcoming.

Dunham, J., I.H. Grant, and S. Watson. *Idealism: The History of a Philosophy*. London: Routledge, 2014.

Guibert, G. *Félix Ravaisson*. Paris: L'Harmattan, 2006.

Janicaud, D. 'Victor Cousin et Ravaisson, Lecteurs de Hegel et Schelling'. *Les Études philosophiques* 4 (1984): 435–50.

Janicaud, D. *Ravaisson et la métaphysique*. Deuxième édition. Paris: Vrin, 1997.

Lodge, P. 'Leibniz on Created Substance and Occasionalism'. In *Locke and Leibniz on Substance*, edited by P. Lodge and T. Stoneham, 186–202. London: Routledge, 2014.

Lovejoy, A.O. 'Some Antecedents of the Philosophy of Bergson'. *Mind* 22, no. 88 (1913): 465–83.

Madinier, G. *Conscience et mouvement*. Paris: Alcan, 1938.

Mauve, C. 'Ravaisson lecteur et interprète de Schelling'. *Romantisme* 88 (1995): 65–74.

Nadler, S. *Spinoza's Ethics*. Cambridge: Cambridge University Press, 2006.

Parodi, D. *La Philosophie contemporaine en France: Essai de classification des doctrines*. Paris: Libraire Félix Alcan, 1920.

Phemister, P. '"All the Time and Everywhere Everything's the Same as Here": The Principle of Uniformity in the Correspondence Between Leibniz and Lady Masham'. In *Leibniz and His Correspondents*, edited by P. Lodge, 193–213. Cambridge: CUP, 2004.

Phemister, P. *Leibniz and the Natural World*. Dordrecht, The Netherlands: Springer, 2005.

Ravaisson, F. *La Philosophie en France au XIXème Siècle*. Paris: Librairie Hachette, [1867] 1895.

Ravaisson, F. 'Philosophie contemporaine'. In *Métaphysique et morale*, edited by F. Ravaisson (1986), 396–427. Paris: Vrin, 1840.

Ravaisson, F. 'Contemporary Philosophy'. Translated by J. Dunham and M. Sinclair. In *Félix Ravaisson: Selected Essays*, edited by M. Sinclair. London: Bloomsbury, forthcoming.

Ravaisson, F., E. Quinet, and F.W.J. Schelling. 'Lettres de Ravaisson, Quinet et Schelling'. *Revue de Métaphysique et de Morale* 43, no. 4 (1936): 487–506.

Risi, V. *Geometry and Monadology: Leibniz's Analysis Situs and Philosophy of Space*. Basel: Birkhäuser, 2007.

Schelling, F. W. J. *Ideas for a Philosophy of Nature*. Translated by E. E. Harris and P. Heath. Cambridge: Cambridge University Press, [1803] 1988.

Schelling, F. W. J. *Jugement de Schelling sur la philosophie de M. Cousin et sur l'état de la philosophie Française et de la philosophie Allemande en général*. Translated and introduced by F. Ravaisson. *Le Cahier* 6: 66–84, [1835] 1988.

Schelling, F. W. J. *First Outline of a System of the Philosophy of Nature*. Translated by K. P. Peterson. New York: SUNY, [1799] 2004.

Simon, J. *Victor Cousin*. Translated by M. B. Anderson and E. P. Anderson. Chicago: McClurg, 1888.

Sinclair, M. 'Ravaisson and the Force of Habit'. *Journal of the History of Philosophy* 49, no. 1 (2011): 65–85.

Vermeren, P. 'Les aventures de la force active en France: Leibniz et Maine de Biran sur la route philosophique menant à l'éclectisme de Victor Cousin'. *Exercices de la patience* (1987): 147–68.

Article

Reviving Spiritualism with Monads: Francisque Bouillier's Impossible Mission (1839–64)

Delphine Antoine-Mahut

This paper studies Francisque Bouillier's contribution to cousinian Spiritualism, from his first text on the History of Cartesian Philosophy from 1839 (revised version from 1842) to the publication of Du principe vital et de l'âme pensante (1864), a work which was likewise considerably amended as a result of the polemics it gave rise to. The paper is concerned with the reception of Leibniz in a double sense. In a positive sense, Bouillier managed to reintegrate in the caricature of the Cartesian soul conceived by the Cousinians, a force that was criticized by the latter. In a negative sense to the extent that, for Bouillier, the direct re-appropriation of Leibniz's dynamic ontology was impossible without completely breaking with Cousin himself. Hence, Bouillier's reception of Leibniz took the form of a progressive suppression of a monadological tendency in favor of a rehabilitation of a theory of minute perceptions. The primacy of the interior sense that results from this allowed him to construct a Descartes who is different from Cousin's but without completely rejecting Cousin. This Descartes is probably closer to Malebranche than to Leibniz. In order, however, to understand this movement in the history of ideas and its lasting echoes in the history of philosophy, a simply study of the reception does not suffice. One must moreover study the importance of a whole series of mediations and prisms that constitute Bouillier's intellectual framework. Here, I have focused on three such prisms : (1) the Cartesian prism itself, which has contributed to the identification of animism as a regression into scholasticism; (2) the Cousinian, prism, which assimilate all resurgence to culpable pantheism ; (3) the prism of the first translations and uses of the polemic between the major figure of Stahl and the way they are refracted in each other. This study is a contribution to a more general reflection on the categories and «labels» that we employ when telling, and telling ourselves, the histories of philosophy.

1. INTRODUCTION

Nineteenth-century French 'spiritualism' is a paradigmatic case of a philosophical movement that is construed and mediated through the reception of early modern authors. The aim of this contribution is to study certain tensions and developments in this movement, focusing on two questions that were formulated by the philosopher Francisque Bouillier. First: Is it possible to overcome the overly narrow Cartesian conception of the soul by appealing to the dynamic ontology of Leibnizian monads? Second: Might the reference to Leibniz help to consolidate rational spiritualism, protecting it from materialist attacks, without however falling into the excesses of those physicians in the medical faculties who adhered to strong forms of animism or vitalism? The timespan studied, from the *Histoire de la philosophie cartésienne* (1839) to the *Du principe vital ou de l'âme pensante* (1864), follows Cousinian eclecticism from its highpoint through its progressive institutional decline. The article combines two analytical approaches. On the one hand, I study the reception of Descartes and Leibniz in Bouillier; on the other, I study how Bouillier's readings had to be mediated through certain positions – Cartesian, Cousinian, and spiritualist-medical – in order to gain an audience. I will thus examine the passages in Bouillier's texts where institutional effects are felt and where he positions himself intellectually in relation to those institutions. At the same time, I reconstruct a philosophical scene that was asymmetrically occupied by authors all motivated by a common effort, namely, mobilizing early modern authors (here essentially Descartes and Leibniz) in order to stand out individually in a stifling institutional context.[1] The analyses will mainly show how Bouillier opened up a new path for spiritualism which, by proposing new readings of Descartes and Leibniz and stressing the notion of an 'inner sense of life', allowed for an authentic dialogue with the life sciences.

2. THE CONTEXT: THE 1839 COMPETITION ON THE HISTORY OF CARTESIANISM

The beginning of the 1840s was a key moment for Cousinian eclecticism. Let us briefly recall who Victor Cousin (1792–867) was, and the all-important role he played in French academia in the nineteenth century. Cousin entered the French Academy and the State Council in 1830. In 1832, he became a member of the Council for Public Instruction where he was in charge of philosophy. He was elected a member of the newly created *Académie des sciences morales et politiques* and appointed director of the *Ecole Normale Supérieure*. In 1840, under the government of Thiers, he was minister of public instruction for some eight months. He presided over the

---

[1] On this, see Vermeren, *Victor Cousin*.

concours d'agrégation – an essential exam for anyone wishing to enter academia – for 27 years. During the period we are concerned with, he possessed all the means necessary to fulfil his intellectual and political ambitions. But for this very reason, he was also very strongly criticized,[2] mainly because his brand of Cartesianism was incapable of providing a satisfactory response to the rise of the life sciences and of responding adequately to the onslaughts against 'pan-metaphysics' from influential figures such as Broussais.[3]

The Cousinian movement was thus, somewhat paradoxically, threatened exactly because it was so powerful. Hence, when the *Academie des sciences morales et politiques* proposed the writing of a detailed history of Cartesianism as the theme of the 1839 competition, the aim was clear, namely exploiting all the resources of Cartesian philosophy in order to strengthen spiritualism against its adversaries. The jury included Cousin, Degérando, Damiron, St. Hilaire, Edwards, and Jouffroy. When the dissertations were submitted (anonymously), they were read by two members of the jury. Their remarks would subsequently be included in a final report, written by Damiron. Winning dissertations were intended for publication, on condition of revision following the suggestions made in the report. In 1839, seven dissertations were submitted: one arrived too late; two received the distinction 'very honorable' (Bouillier and Renouvier); Bordas-Demoulin won the prize. Among the remaining three contestants, the identity of only one was ever divulged: Jean-André Rochoux, a somewhat subversive nurse working at the Hospital Bicêtre. Rochoux published his dissertation at his own expense in 1843 under the title *Épicure opposé à Descartes*.[4]

Apart from the simple servile reproduction of the Cousin's doctrine, the candidates had the choice between the following two approaches. The first was to attack Cartesianism frontally, attempting to completely overturn the philosophical landscape. This was the approach chosen by Rochoux and obviously banned him immediately from the podium of the winners. The second approach, which might be rewarded and result in the dissertation's publication, on condition of revision, would be to find ways to revise spiritualism by means of theoretical tools found outside the official doctrine of Descartes, while still remaining within the sphere of Cartesianism. Here, I study the dissertation and subsequent philosophical development written by a candidate of the second variety, namely Francisque Bouillier (1813–99). Bouillier was situated at the heart of the French Academic institution throughout his entire career. He was admitted to the *École Normale*

---

[2]On spiritualism in nineteenth century France, see Janicaud, *Ravaisson et la métaphysique*.
[3]François Joseph Victor Broussais (1772–838) was a student of Bichat and Pinel, head of the *Hôpital Val de grâce* in Paris, professor at the medical faculty and inspector of public health. He was a member of the *Académie de médecine* founded in 1823 and of the *Académie des sciences morales et politiques* from 1832 onwards. He wrote extensively on the relation between the history of philosophy and the history of medicine. He had an influence comparable to Cousin's whom he was constantly challenging in public.
[4]On this, see Antoine-Mahut, 'Cartésianisme dominant et cartésianismes subversifs'.

Supérieure in 1837. He obtained his *agrégation* in 1837 and doctorate in 1839. He was professor at the *Collège d'Orléans* in 1837–9, then at the Faculty of Letters in Lyon, where he was the dean from 1849 to 1864. He was rector of the Academy of Clermont-Ferrand in 1864 and general inspector of public instruction from 1864 to 1867. From 1871 to 1876, he directed the *École Normale Supérieure*. He was elected to the *Académie des sciences morales et politiques* in 1875 and became its president in 1889.

Rochoux summarizes very neatly what Bouillier's intentions were when submitting a dissertation for the 1839 competition:

> After reporting the grueling judgment, the Ferney patriarch passes on Cartesianism (*Dict. Philosophique*), Mr. Libri recalls how this philosophy in our days found some eloquent defenders; but he sees in this 'a reaction that seems excessive and consequently transient' (*Revue des deux mondes*, September 1842, 742). It will not be long before his prediction is fulfilled, or rather, it already has been, according to Mr. Francisque Bouillier, one of the winners of the prize on Cartesianism. This is how this author finishes his long in-8° dedicated to the critical examination of Descartes's philosophy: 'As a system of philosophy, *Cartesianism is dead*, but it has left deep traces of its passage in the sciences, for it has given nineteenth-century French philosophy its method and some of its principal results. Cartesianism is dead, but its spirit lives in us, it is the very spirit of science, philosophy and modern civilization' (*Histoire et critique de la revolution cartésienne*, 442–3). Considering that a dead system lives on in spirit is one of those ingenious feats that only spiritualism is capable of. They have nothing to learn from Roland, who said of his famous mare: 'Indeed, she is dead, but she has no other flaw I can think of'.
> (Arioste, Roland furieux, t. II, c. XXX, 426, in Rochoux, *Epicure opposé à Descartes*, 112–113, note 2)

For Bouillier, reviving the 'spirit' of Cartesianism involved careful work on the central figures of the Cartesian philosophy but wholly contextualized within institutionalized Cousinian spiritualism.

In his *Histoire de la philosophie cartésienne*, Bouillier presented a duel between the official Descartes and the Leibniz of the *Monadology*. This duel was, however, constantly undercut by a subtler, dialogical relation between the official Descartes, the Leibniz of the *Monadology*, and the Leibniz of minute perceptions. In order to understand how and why, we must read Bouillier's later work, including his generally overlooked text from 1862, the *Du principe vital et de l'âme pensante, ou Examen des diverses doctrines médicales et psychologiques sur les rapports de l'âme et de la vie*.[5] This last book should be understood in the context of a

---

[5] This text is the revised and considerably augmented version (420 pages) of a dissertation of about 60 pages submitted in 1858 titled *De l'unité de l'âme pensante et du principe vital.*

controversy between, on the one hand, the heirs of Maine de Biran[6] and certain spiritualist physicians reviving animism, and, on the other, the Cousinians who were staunchly opposed to anything that had even a remote resemblance to pantheism. But in a context where King Cousin was dethroned and aging, should we interpret Bouillier's 'third way' as a concession to, or an attack against, those adversaries of Cousinianism?

3. THE HISTOIRE DE LA *PHILOSOPHIE CARTÉSIENNE* (1839–54)

The *Histoire de la philosophie cartésienne* is a work that a great many Descartes scholars continue to refer to today. The work on the vital principle, in contrast, has been largely forgotten. In order to highlight the most significant changes between the *Histoire de la philosophie cartésienne* and the text on the vital principle, I here refer to the second, amended edition of Bouilliers's dissertation, published in 1842 under the title *Histoire et critique de la revolution cartésienne*. Incidentally, this is also the text that Rochoux refers to. I will consider in what respects the Cousinian institutional and conceptual framework constrained Bouillier's attempts at becoming more autonomous and the concrete effects this had in his texts. I thus show how the *logic* of Bouillier's philosophical thought is tightly linked to the *history* of his philosophical thought.

When rewriting his *Histoire de la philosophie cartésienne* in view of publication, Bouillier found himself in a double-bind. On the one hand, he had to yield to the pressure of Victor Cousin and Jean-Philibert Damiron[7] who both required that he put more stress on the metaphysical questions and their

---

[6]On the return of Maine de Biran's doctrine and of Leibniz's dynamical doctrine on the scene of the French universities from 1840 onward, see Janet, 'Une nouvelle phase de la philosophie spiritualiste', 365–9. Janet includes in this movement Jouffroy, Ravaisson, Vacherot and even Saisset. Biran died in 1824. It should be recalled here that Lainé, a close friend of Biran and the executor of his testament, put Cousin in charge of examining and editing the papers he had in his possession. Among these documents, Cousin had the *Examen des leçons de philosophie de M. Laromiguière* (which had already been published in 1817), the 1819 text on Leibniz (*Exposition de la doctrine philosophique de Leibniz*), and the manuscript *Nouvelles considérations sur les rapports du physique et du moral*, a work that had received a prize from the Academy of Copenhagen and the 'great importance' of which Cousin stressed when publishing it for the first time in 1834 along with the other two texts. Cousin was accused of having deliberately delayed (for more than ten years!) the publication of the papers he had in his possession and of having censured them when he did finally publish them. The affair is recounted in Leroux, 'De la mutilation d'un écrit posthume de M. Jouffroy', 293. This heavy context weighed upon all attempts at rehabilitating Biran's duo-dynamism, that it is to say, his theory regarding the two sources of vital phenomena: the active I and the unknown vital, animal force.

[7]Jean-Philibert Damiron (1794–862) became a member of the *Académie des sciences morales et politiques* in 1836 and obtained the chair in modern history at the Sorbonne in 1838. Victor Cousin put him in charge of writing up the reports on the dissertations presented at the 1839 competition. Damiron was the author of an *Essai d'histoire de la philosophie en France au dix-neuvième siècle* (1828) and an *Essai sur l'histoire de la philosophie en France au dix-*

foundational character in the Cartesian tree of knowledge. Consequently, he should stress the distinction between the soul and the body.[8] On the other, Bouillier himself wanted to point out those things which, in the institutionally dominant representation or image of Descartes, remained too 'abstract' or too 'pure' for him to give right answers to the 'empirical philosophers'. Bouillier was thus tempted to stress the union of soul and the body while diminishing the passive dimension of the soul. But yielding to that temptation would oblige him to reattribute a kind of activity to the Cartesian soul. However, after having, at first, stressed spontaneity as being at the heart of causation, the latter being inseparable from the notion of substance, Cousin now described this notion of activity as leading on a slippery slope towards pantheism since it 'gathers all the material and spiritual phenomena in a single, simple and indivisible subject in which all the diverse laws of nature are fulfilled'.[9]

What results from this intellectual tension in Bouillier's text is a reading of Descartes that, at the same time, presents him as the best dualist metaphysician of all and also as he who neglected the activity or force of the soul. For a significant example, let us consult the following passage:

> One must reproach Descartes for having made the notion of spirituality foreign to the notion of force, for having underestimated the essential activity of the soul, and for having defined it by its action rather than by its essence when defining it by thought alone. But we cannot grasp thinking without a subject, nor can we grasp any of the soul's phenomena without an active principle, without a force producing them.
> (Bouillier, *Histoire de la philosophie cartésienne*, 65–6)

This problem expands into a general underestimation of created substances that puts Descartes on the slippery slope towards the phenomenalization of the soul's thoughts. Consequently, Descartes was left in the mercy of those physicians among Bouillier's contemporaries who also took an interest

---

septième siècle (1846). His report on the dissertations on Cartesianism was published in the introduction of the second book.

[8]For details regarding the revisions required by Damiron, see Antoine-Mahut, 'La fabrique de l'histoire du cartésianisme néerlandais dans les histoires de la philosophie française du dix-neuvième siècle' et 3: With regard to Bouillier, Damiron cleverly stresses a proximity between Locke and Descartes, in relation to the theory of the soul's passivity (see Damiron, *Essai sur l'histoire de la philosophie en France au XVIIe siècle*, 50–2.)

[9]Cousin, *Cours d'histoire de la philosophie moderne*, 11e Leçon, 88. On the evolution of Cousin's position on pantheism from his first classes from 1815 to their reformulation in the *Histoire générale de la philosophie depuis les temps les plus reculés jusqu'à la fin du XVIIIe siècle* (1863), see Moreau, 'Spinoza et Victor Cousin', 327–31; 'Spinozisme et matérialisme au XXe siècle', 85–94; 'Saisset lecteur de Spinoza', 85–97; 'Trois polémiques contre Victor Cousin', 542–8; 'Les enjeux de la publication en France des papiers de Leibniz sur Spinoza', 215–32; Vermeren, 'La philosophie au présent', 167–80; Macherey, 'Leroux dans la querelle sur le panthéisme', 215–16; Cotten, 'Spinoza et Victor Cousin', 231–42.

in philosophy in its official and institutional dimension, and who wanted, either to simply get rid of Descartes in order to establish a kind of materialist monism, or use mechanism in order to transform Descartes himself into a materialist. Phenomenalizing the soul's thoughts consists in considering them independently of their substratum. And from considering them as independent from this substratum to completely removing the latter, there is only a short step – a step that leads from empirical psychology to monist materialism. The dangers that Cousin had underlined regarding pantheism showed up in Descartes himself. It thus appears as if, by reason of the institutional hold that Cousin's 'passive' Descartes had in France, attributing activity to a soul united with the body would require getting rid of Descartes altogether. Or said in other words: granting activity to the Cartesian soul would amount to a disavowal of Cousin himself.

Such a conflict appears very clearly in the treatment Bouillier reserves for a text much underestimated by Cousin, the treatise on the *Passions de l'âme* (Bouillier, *Histoire de la philosophie cartésienne*, volume I, chap. V, livre I).

After making the soul pure thinking, Descartes should have understood the passions as pure phenomena of the body, but since he did not dare to be that rigorous, he limited himself to making the soul's contributions as small as possible, and to relating them almost exclusively to the body and the animal souls, the great material agent by means of which he explains their generation and all their movements (Bouillier, *Histoire de la philosophie cartésienne*, 111–12).

This somatic interpretation of the passions as a unitary phenomenon culminates, in the second part of the *Histoire*, in a partial reading of Descartes's enumeration of the passions. While recognizing the importance of the principal 'ways' in which the *soul* relates to an exterior object in the phenomenon of passion, Bouillier maintains that the criterion's distinguishing phenomenon resides in the 'impression' that these objects make on the *brain*. Instead of being placed on the side of a *soul united to a body*, the question of the passions' 'primacy' thus remains on the side of the *body alone*:

> […] it is in the temperament of the body, in the impressions on the brain, and in the objects that can act upon our senses that he places the most ordinary and most general cause of all our passions. We must thus examine these objects and the various impressions that they can make upon us, in order to arrive at the most complete classification possible of the soul's passions. Even though they are innumerable, they only affect us in a certain number of ways, following whether they hurt us or are profitable for us. For there are only as many primary and primitive passions, as there are basic ways in which our senses can be moved by an object.
> (Bouillier, *Histoire de la philosophie cartésienne*, 112)

What the Cousinian framework prevents Bouillier from stressing, and which in turn justifies the contradictions and insufficiencies that he imputes on Descartes, is thus the inner sense specific to the passions *of the soul*. The impossibility of formulating such passions becomes manifest once again towards the end of the chapter, where Bouillier establishes an opposition between the *passions* of the body and the *emotions* of the soul:

> To the passions of the body must be opposed the inner emotions or passions of the soul, that is to say, those passions that are aroused in the soul by the soul itself and not by some movement of the animal spirits. Happiness and misery in our lives depend on these emotions in particular, since they touch us more closely. Often they are joined with corporeal passions that correspond to them, but equally often they meet with those that are contrary to them, and which serve to counterbalance them.
> (Bouillier, *Histoire de la philosophie cartésienne*, 117)

One ultimate and ingenious way of justifying such an approach, while still maintaining the pertinence of the dualist interpretation that Cousin inflicts upon the recalcitrant Cartesian texts, consists in suggesting that this was the solution Descartes himself proposed in relation to the scholastics. Cousin's obsessive fear of pantheism would then be nothing but the counterpart of Descartes's struggle against hylomorphism. And this double-bind, between pantheism and hylomorphism, makes it impossible to conceptualize 'the true nature of the animal' within the Cartesian framework alone.

> [...] we should take care to notice that this hypothesis regarding the animal-machine stands in a close relation to the rest of Descartes's philosophy. It is not only tied to the totality of his physiology, but to the fundamental principles of his metaphysics. After having fixed the human soul exclusively in its conscious thought of itself, outside of which there is nothing but material and inert extension, after having deprived the creatures of all force and all causation and turned God into the sole force and efficient cause, where could he possibly find the principles of life and sentiment to animate the beasts? Since, according to his system, he cannot possibly give them any active force, he ought necessarily to conceive of them as pure inert matter, subjected to the general laws of movement. This is the link between the automatism of beasts and the metaphysics of Descartes. The true nature of the animal must be reestablished against Descartes.
> (Bouillier, *Histoire de la philosophie cartésienne*, 156–7)

The constraints that the Cousinian framework imposed on Bouillier's reception of Descartes had two consequences. On the one hand, Descartes was pulled in the direction of occasionalism, depriving creatures of all activity. On the other, Bouillier finds himself forced to turn to Leibniz, and more precisely to the *Monadology*, in order to conceptualize this lack of activity: 'It is by means of Leibniz that we will correct Descartes' (Bouillier, *Histoire de la philosophie cartésienne*, 136).

The difficulty at this point, in 1854, is claiming a Leibnizian heritage while avoiding completely dismissing Descartes. He must satisfy Cousin who was in the middle of revising his *Fragments de philosophie cartésienne* and did not overlook any of the successive developments of Bouillier's dissertation. On this point, Bouillier prudently opts for a quantifying formula that leaves in place a small margin for 'reform':

> Doubtless one is allowed to give this designation [i.e. reformed Cartesianism] to a philosophy about which it can be said, with Mr. Cousin, that is three-quarters that of Descartes.
> (Bouillier, *Histoire de la philosophie cartésienne*, 405)[10]

But what does the remaining quarter consist in? A Leibnizian reform of the notion of substance. Leibniz reintegrates force and activity into the essence of substance without however going back to 'the naked force of scholasticism' (Bouillier, *Histoire de la philosophie cartésienne*, 411). By replacing substantial extension with monads he, 'finally', replaces the outdated notions of the scholastics that Descartes had been right to reject. But he also rejects something that, when described in these terms, can no longer be described as a merely anecdotal 'quarter', namely Descartes's own metaphysics. Bouillier uses a detour via occasionalism to describe this metaphysics in the very same terms as those Cousin used to describe the dangers of pantheism and Spinozism:

> By correcting the fundamental error of Cartesianism, he also finds the remedy for Spinozism. Here we finally have real and persistent substances that are as different from the phenomena as they are from the infinite substance itself, and not just ghostly substances without consistence or action that are absorbed into the true cause and the unique true substance. If there are no such substances, there are no obstacles to pantheism. Without monads, as Leibniz says it very well, Spinoza would be right.
> (Bouillier, *Histoire de la philosophie cartésienne*, 412)

Bouillier reproached Descartes for thinking man as a soul distinct from a body with which no real, effective synergy is possible. To this, he now opposed a Leibniz who, by way of reaction, reaffirmed the 'indissoluble union of the soul with its organs' (Bouillier, *Histoire de la philosophie cartésienne*, 433). In Leibniz, however, the theory of pre-established harmony had the disadvantage of reintroducing, within Leibniz's own system, a principle of passivity that pulls Leibniz towards occasionalism and runs counter to the principles of the monadology: 'There is thus a true contradiction between the essential principles of monadology and pre-established harmony, in such a way that it seems necessary to sacrifice pre-established

---

[10]Bouillier develops his general analysis of Leibniz in volume II (chapters 17–19 are explicitly dedicated to Leibniz).

harmony in order to save the monad' (Bouillier, *Histoire de la philosophie cartésienne*, 473). When defending the notion of pre-established harmony, Leibniz is as close to a Descartes incapable of conceptualizing action as he is to a Malebranche inclining towards Spinoza. What is lacking in both occasionalism – including Cartesian occasionalism – and pre-established harmony is a veritable notion of activity.

In 1854, the winner of the competition thus found himself in an impasse. Descartes was too much of a support for the system he was fighting. But if the very definition of Cartesianism would have to be changed, and if official Cartesianism was too dominant to make such change possible, maybe it was time to kill Rochoux's mare and renounce on Cartesianism altogether? And if there was truth be found on the side of Leibniz, maybe one should in fact return more decisively to him. But this would have to be done in such a way that it allowed the conceptualization of a new animism without any risk of materialism and pantheism, and consequently by downplaying if not outright rejecting the passivity reintroduced by the Leibnizian conception of pre-established harmony. In the *Histoire de la philosophie cartésienne*, Bouillier only hinted at his solution to the problem once. Hence, while speaking of Leibniz and of some future project not yet realized, he upholds the ideal of a 'doctrine of the unitary principle of life and thought' (Bouillier, *Histoire de la philosophie cartésienne*, 422). But what this referred to was not a simple return to Leibniz. It was rather a personal position that Bouillier adopted in a contemporary intellectual debate about animism. Leibniz, in other words, was nothing but a means to intervene in this debate.

4. THE RETURN OF ANIMISM ON THE PHILOSOPHICAL AND MEDICAL SCENE (1850–60)

In the context of the debates on animism, it is important to note that most participants claimed Cousin's authority, or that they all attempted to align Cousin's position with their own. As an example, I take the discussions around the *Introduction à la philosophie médicale* by Richard de Laprade. This dissertation, presented to the Academy of Lyon in July 1860 shortly before Laprade's death and subsequently published by his son under the title *Animisme et vitalisme*,[11] defends the animist position adopted by the physicians of the Montpellier School against Bouillier. He described the commotion Bouillier had occasioned among organicist physicians, even among those who were supporters of the Cousinian school. The organicists were those physicians, spiritualist or not, who believed that the consideration of organic life should be put at the heart of the reflection on the

---

[11]Laprade's dissertation was published in Lyon by Aimé Vingtrinier and includes only 16 pages.

nature of the soul. He also cited statements by Cousin seemingly going in the direction of animism and organicism:

> Mr. Cousin's opinion is even more explicit: 'My conviction is that under these different organs there is a force that makes them act and concur towards life; a force that, when the exercise of the functions has been upset, will intervene by means of external stimuli and internal affections and reestablishes more or less the harmony among the functions, or even the workings of each function. At all times, some force has been admitted, be it obscurely and by means of designations that were more or less precise, such as Plato's *appetitive soul*, Aristotle's *sensitive soul*, or the *conserving* or *mediating soul* evoked by certain more modern physiologists. It is a force that cannot be denied, on pains of falling into the kind of crude materialism that only sees organs in the body, or loosing oneself in a subtle and chimerical spiritualism that confounds the vital principle with the principle of spiritual life itself.
> (Laprade, *Animisme et vitalisme*, 4)

Responding to Laprade in the 1862 version of his dissertation, Bouillier blew his own Cousinian cover. First, Bouillier argued, Laprade reported only oral statements[12] that were published nowhere by Cousin. There was thus nothing to guarantee their authenticity. Secondly, there was surely one text where Cousin praised animism, namely in his edition of the *Phaedo*. But this remained indirect praise stated in the context of the examination of a philosophy other than his own, namely Platonism. Consequently, there were concerns about intellectual honesty in relation to Laprade's reading, warranting the greatest prudence in relation to his account of Cousin's position: 'It is more exact to say that, to our regret, Mr. Cousin has not treated in any particular way this question on which he could have enlightened us, as he has on a great many others' (Bouillier, *Du principe vital et de l'âme pensante*, 297).[13] In fact, Cousin had discussed animism with Bouillier on a prior occasion. But the replies Cousin had given did not really have a bearing on the truth of the doctrine. They highlighted the 'advantages' of the thesis, namely its greater ability to ward off the materialist enemy than Maine de Biran's brand of animism. Cousin's reply to Bouillier was thus situated on a *strategic* terrain and did not concern his *personal* position. But importantly it did not indicate how to take into account the organicist position:

> I must admit that your opinion regarding the identification of the principle of life and thought does not appeal to me, accustomed as I am to judge the

---

[12]What Laprade had in mind was probably the unpublished seminars on consciousness from 1819 to 1820. They can be consulted in the manuscript collection of Cousin's writings at the library of the Sorbonne.

[13]Regarding the argument from the *Paedon*, in connection with Stahl, see Bouilllier, *Du principe vital et e l'âme pensante*, chap. V, 73.

principles by the phenomena which are here entirely different. Moreover, I do not really see the advantages of this delicate discussion for the present state of philosophy. It is important that we focus our forces on the points under threat, including the spirituality of the soul and the essential distinction between God and the world.
(Latreille, *Francisque Bouillier*, 226–7. The letter is dated 15 November 1859)

Bouillier responded to Cousin in the very first line of the new preface of *Du principe vital et de l'âme pensante*:

[...] how temerarious it is pretending to overturn the doctrine of the soul's duality and of life established so solidly by Maine de Biran and by Jouffroy within spiritualist philosophy and wanting to rehabilitate – and this not without risks for the spirituality of the soul – the universally rejected animist hypothesis! How imprudent, in these times, to present to sceptics and materialists the pleasant spectacle of serious dissidence among the spiritualist philosophers! We reply, first of all, that it is a question of truth and not tactics; and besides, it is my firm conviction that the doctrine I would like to triumph would result in the strengthening and not the downfall of spiritualism.
(VI–VII)[14]

Bouillier's objective in the second revised version of the book on the vital principle was thus to show the way in which animism, while promoting a doctrine that unified the soul and the vital principle, allowed him to respond to the organicists, both spiritualist and non-spiritualist, better than the followers of Biran could do.

5. COUSINIAN CONSTRAINTS ON THE *DU PRINCIPE VITAL ET DE L'ÂME PENSANTE*

Bouillier's classes at the Faculty of Letters in Lyon – classes that formed the basis for his last publications – were dedicated to this question regarding the unity of the vital principle and the thinking soul. He presented a first summary of this work to the Académie des sciences morales et politiques, and then published it in 1858. Subsequently, he continued to work on it, after it provoked heated discussions not only with the medical schools and in particular with the organicists, but also within his own spiritualist camp.

The terms of the debate thus changed between the history of Cartesian philosophy and the dissertation on the vital principle. While Cousin placed Bouillier in an antagonistic position in relation to the physicians, Bouillier now came to share terrain with them. The spiritualist reflection

---

[14] The original version of the dissertation even spoke of this as a 'family quarrel' (see -1858- 5). Hence, corresponding with Cousin brought Bouillier to reconsider or least render more explicit the relation between the 'family quarrel' and the 'battle against materialism'.

on the soul did not exclude *a priori* every possible organicist consideration. In order to negotiate that terrain theoretically, Bouillier constantly produced bifurcations in theories that Cousin presented as uniform. He thus has a double Descartes, a double Leibniz, a double spiritualism, a double organicism, etc. Each of these Descartes, Leibniz, spiritualism, organicism, reflect and refract the constraints of the institutions that he takes part in. These multiple bifurcations are characteristic and original in Bouillier's readings, making him stand out as someone who always proposed a 'third way'. One can identify three discursive, conceptual or even political strategies, aiming at cleverly circumventing the Cousinian constraints. I here list them in an order going from the most polemical to the most conciliatory: (1) A strategy that inverts the Cousinian project itself and which consists in relating the defects of Cartesianism to abusive organicism. (2) A strategy that consists in showing that, in Descartes, those defects stem from the firm stance he had to adopt in the polemics opposing him to the scholastics. (3) Finally, a strategy that consists in associating himself with the most orthodox among the Cousinians, namely Damiron. Let us consider these three strategies one by one.

The strategy of inversion. The text on the vital principle is original in relation to the dissertation on Descartes in that it makes use of an argument, found among most empiricist thinkers from the time of Descartes and the physiologists of the nineteenth century, according to which Cartesian mechanism leaves vacant a whole theoretical area that needs explaining, namely that of vital phenomena. Among Descartes's contemporaries, Bouillier in particular refers to a letter where Descartes comments on the first physiological theses of the Utrecht physician Regius, reproaching him for not having gone far enough in his mechanical explanation of life and having left 'a greater difference between living and non-living things than between a clock and any other automaton' (For the texts by Descartes quoted here, see Victor Cousin, *Oeuvres de Descartes* (Paris, 1824–6) vol. 8, 628). This passage is emblematic of the Cartesian excess, but it also indicated that mechanism had to be corrected or completed when it came to vital phenomena, including for an empiricist like Regius. As for what concerns the French physicians and physiologists of the nineteenth century, none of them claimed to explain the organic functions solely by means of the laws of mechanics. And even if one appealed to the basic principle of Cartesian physiology – that is, 'fire without light' following the expression from the *Traité de l'Homme* – as the constitutive element of a larger theory of the living, the following point remained patently clear: 'Among the various fires of nature that might be called upon to replace the vital soul, Descartes made an unfortunate choice, namely carbonic gas which is a principle of death, and not of life, as modern science has shown' (Bouillier, *Histoire de la philosophie cartésienne*, 175). In the revised version of *Du principe vital*, Bouillier attempted to pull Descartes towards the position of those physicians responsible for the progressive sidetracking of the soul in the constitution of the science of man, emptying Descartes's doctrine of its spiritualist potential. Bouillier thus literally turned Cousin's project on its head: The interest of Cousin's Descartes became the contrary of what it was for Cousin.

The strategy of excess. In order to be equitable towards the Cartesian project, however, it had to be stressed that Descartes had been forced to harden his arguments in order to fight the pernicious effects of a ruinous kind of animism, coming at much too high a cost for humankind. This 'overflow of souls, in man and in nature' was emblematically expressed in the doctrine of Jan Baptist Van Helmont, whose doctrine is judged as follows:

> There can be no doubt that this prodigious life of souls and intellectual faculties has contributed, along with the general principles of his metaphysics, to throw Descartes into the opposite exaggeration of mechanical physiology, the absolute negation of life and of all souls except the human ones, thus endorsing the automatism of beasts.
> (Bouillier, *Histoire de la philosophie cartésienne*, 164)

But this Cartesian excess had an unfortunate effect: while animism was the prevalent system in the history of ideas before Descartes, Cartesianism had left such a powerful imprint on posterity that the very term 'animism' could no longer be pronounced without immediately lending to a 'suspicion of ontology and a return to scholasticism' (Bouillier, *Histoire de la philosophie cartésienne*, 48). What Bouillier had to accomplish next, then, was not so much the redefining of animism, but the refutation of that with which it had generally been identified with in the Cartesian history of ideas, by partisans as well as opponents of those ideas. Descartes, but also Cousin reading Descartes, had contributed to the exaggerations attached to the term 'animism', but without such exaggeration being strategically ungrounded.

The strategy of appending. In order to balance his project, Bouillier took a final precaution which was strategically very clever: He identified an ally among the Cousinians who had tried to stay in the background: Jean-Philibert Damiron, and in particular his *Rapport sur le concours sur la philosophie de Leibniz*, published in 1860. It was Damiron who had asked Bouillier to reorient his first work towards Descartes the metaphysician. In his own *Essai sur la philosophie en France au XVIIe siècle* (1846), he had expressed scruples about Leibniz. These scruples were expressed at the time when Leibniz's philosophy had been chosen as the topic for the essay competition of the *Académie*. This in itself testified to the actuality current relevance and the importance of the Hanover philosopher. Influenced by the dissertations written by Nourrisson and Foucher de Careil,[15] Damiron underlined what,

---

[15]Cousin's 'verbal report' on this dissertation for the *Académie des sciences morales et politiques* dates from 1854. It was published in *Leibniz, Descartes et Spinoza. Avec un Rapport par M. V. Cousin*, an augmented version of Foucher de Careil's, *Réfutation inédite de*

for him, constituted the most pertinent aspect of Leibniz's dynamics, namely the way it reintroduced a notion of force that was neither materialist nor 'purely' spiritualist. Leibniz, 'who does not always tell us all he does or how he does it', had drawn his argumentation from the science of the soul, not from his conception of God or matter (Bouillier, *Histoire de la philosophie cartésienne*, 48).[16] This aspect of Damiron's reading of Leibniz was crucial for Bouillier, but it also helps to clarify the kind of obstacle Bouillier was facing specifically when revising his dissertation on the vital principle. This obstacle concerned not only Cartesianism or animism, but the complex relation between the two theories that stand out when they are refracted in the prism of the institutionally dominant Cartesianism: 'The false notion of the soul endorsed by Descartes [we must here understand: by Cousin's Descartes]: that is the principal cause for the resistance that animism today encounters in French philosophy' (Bouillier, *Histoire de la philosophie cartésienne*, 319).

Finally, we must consider one of the most important readings of Leibniz at the time, namely the one proposed by Albert Lemoine who read Leibniz through the eyes of Georg Ernst Stahl, a pietist physician who had engaged in an acrid controversy with Leibniz on the question of vitalism and pre-established harmony, subsequently published by Stahl in 1720 under the title *Negotium otiosum* ('A Tiresome Quarrel'). Once we understand the interest of Lemoine's text on Stahl for Bouillier, we can better grasp the interpretation and rehabilitation of Leibniz that he proposes as a response to Lemoine without however ever losing sight of Cousin.

In France, Stahl had become central to contemporary philosophico-medical debates, and the way he was read is telling of the confusion that ruled when it came to the meaning of the term 'animism'. For certain commentators, Stahl had simply replaced Jan Baptist Van Helmont's *archeus* with the rational soul. For others, he was mainly influenced by Descartes, Malebranche and, more generally, by doctrines about the passivity of matter. Indeed, Stahl's work represented a kind of indeterminate polemical reference that everyone gave a meaning to according to their needs. Probably for this very reason, Stahl was at the heart of the academic debates. Maine de Biran was the first to recognize his importance.[17] After him, Stahl was talked

---

Spinoza par Leibniz. Nourrisson's, *La Philosophie de Leibniz* was published with a quotation attributed to Cousin on the cover: 'Leibniz is the last and greatest Cartesian'.

[16] Bouillier refers to Leibniz as an ally from very early on, referencing Leibniz's *De ipsa vi naturae insita,* included in the Erdmann edition: '[...] the first science of force is not truly mechanics or physics but, as Mr. Damiron has pointed out [in the report for the *Académie des sciences morales et politiques* regarding the competition on Leibniz's philosophy], but psychology' (*De l'unité de l'âme pensante et du principe vital*, 5).

[17] See Biran, *Exposition de la doctrine philosophique de Leibniz*. This text had been written between April and July 1819 and was intended for Michaud's *Biographie universelle*. Biran describes Stahl's physiology as the key theory that would allow for the alleviation of the indetermination of the Cartesian doctrine of innate knowledge and the reconciliation of

about in Paris. Bouillier notes, for example, the influence of the doctoral thesis by Ernest-Charles Lasègue from 1846, *De Stahl et de sa doctrine médicale*. At the same time, Stahl's work was translated in Montpellier.

The most important vindication of Stahl was, however, proposed by Albert Lemoine, a professor of philosophy at the Faculty of Letters in Bordeaux, in his dissertation *Stahl et l'animisme* of which the Academy published a summary in July and August 1858. This dissertation was doubly important for Bouillier.[18] First, it provided him with the conceptual means to deconstruct the modern, pejorative understanding of animism. Next, it obliged him to revise his own position in the book on *Le principe vital*, taking into account the objections that Lemoine made to him in a second version of his work, published in 1864 under the title *Le Vitalisme et l'animisme de Stahl*.[19] The difference between the titles of the two versions of Lemoine's text alone testifies to what was at stake. In the first version, Lemoine's aim was to justify Stahl and revise the meaning of the 'label' of animism. In the second version, on the contrary, he wanted to situate himself in relation to animism, and in particular question the relation it upholds to vitalism. These discussions of Stahl thus served to outline the contours of a new animism that Lemoine had contributed to rehabilitate without, however, subscribing to it, but that Bouillier now wanted to see become dominant within institutionalized spiritualism.[20]

From there on we can distinguish yet another two strategies. First, Bouillier found arguments in favour of animism in Lemoine's dissertation, or rather an alternative way of dissociating animist explanations from the pejorative image that had haunted them since Descartes. Next, he employed and modified those arguments for his own purposes by engaging in a deeper enquiry into Leibniz, in particular on the question of insensible perceptions. Let us take a look at each of these two points.

The importance of Lemoine's dissertation for Bouillier was, first, that it put into focus a kind of double interpretive manoeuvre the effect of which had been to deprive Stahl of all originality: While the Cartesian reception of animism had reduced it to the caricatures of Van Helmont, Parcelsus, Fludd and Boehme, the Leibnizian reception of Stahl had reduced his animism to materialism. Stahl, however, had in reality proposed a true alternative (Lemoine calls it 'an intermediary system') to both mechanism and dynamics. His animism could in this sense be considered radically new, in philosophy as well as in physiology. In order to show that, one

---

Descartes and Locke (see, for example, Exposition de la doctrine philosophique de Leibniz, 20).

[18]Bouillier summarized his debt to Lemoine as follows: 'I have made use of the excellent analysis of this polemics that Mr. Lemoine has provided in his dissertation on Stahl' (Du principe vital et de l'âme pensante, 204).

[19]Lemoine, *L'Animisme et le vitalisme de Stahl*.

[20]Cousin himself called for such a confrontation (see Cousin to Bouillier, 8 November 1858, in Latreille, *Francisque Bouillier*, 226).

would however first have to dissociate Stahl from Descartes. More precisely, one had to dissociate him from a dualist and occasionalist Descartes, a Descartes who had championed the passivity of both the soul and the body. This would permit granting the soul – and only the soul – a certain dynamism, by way of juxtaposing Stahl's position to the doctrine of pre-established harmony and its shortcomings, stressed by Bouillier in his 1854 *Histoire*. Next, Stahl's animism should be rationalized, by being correlated with experimental results and with the struggle against abstract hypotheses. In this context, attributing life to the rational soul became the natural consequence of an unwillingness to multiply secondary and chimerical causes, *archeuses* and genies of all kinds. In summary, Lemoine incited his readers to revise the 'labels' used by the nascent historiography of the life sciences and attempted to liberate the field of animism for this new science.

In chapter VIII of *Du principe vital et de l'âme pensante*, Bouillier analyses the major differences between Stahl's spiritualism and 'that of Descartes and the psychologists who separate the organic functions from thought and will in order to refer them to another principle' (Bouillier, *Du principe vital et de l'âme pensante*, chap. VIII, 152). He complemented this analysis with a polemical study of Leibniz. Lemoine had questioned Leibniz's role in the identification of Stahl as a 'materialist', and thus in the association of Stahl to the 'thousand-headed' monster that Cousin was chasing. Lemoine's aim, however, was to show that if Leibniz did indeed attack Stahl, he did not reject him entirely. When attacking Stahl, Leibniz only aimed at exhibiting those doctrinal elements which, *in an animism Leibniz otherwise subscribed to*, contradicted the theory of pre-established harmony. Moreover, instead of only criticizing the idea according to which the soul directs not only voluntary movement but also the organic functions, Leibniz made the mistake of attacking generally the idea of the interaction between body and soul. He mistakenly took all notions of soul-body interaction – a notion that is not Cartesian if one insists upon making Descartes an occasionalist – to be materialist. At the same time, Leibniz had reinforced the element of passivity in pre-established harmony, hereby compromising the monadology.[21] In sum, by insisting on pre-established harmony, Leibniz lost what the monadology had gained him, namely the activity of the soul. Stahl, on the contrary, used that gain and brought out its full potential:

> Stahl is indeed a true [...] spiritualist, but, reflecting with his common sense along with the rest of the world, he thinks that the spiritual soul is nevertheless capable of moving the body. This is the crime that the author of the pre-established harmony cannot forgive him.
> (Bouillier, *Histoire de la philosophie cartésienne*, 163)

---

[21] For Lemoine's very convoluted argumentation, see *Stahl et l'animisme*, 147–51.

Hence, according to Lemoine, a well-conceived animism did not give rise to materialism. Quite to the contrary, it had as its condition *sine qua non* the immateriality of the soul. Stahl 'brings spiritualism down to the level of medicine, to the advantage of medicine and of philosophy itself' (Bouillier, *Histoire de la philosophie cartésienne*, 187). He went further than Leibniz who remained Cartesian or occasionalist, according to the typology established by Bouillier in his *Histoire*.

The importance of Lemoine's dissertation for Bouillier's project should now be clear. But Bouillier still had another two problems to deal with. First, he had to displace the main foyer of animism from medicine to the history of philosophy. This required downplaying the element that Lemoine had shown to play an overwhelmingly strong role, namely Cartesian mechanism, which had helped legitimize materialist re-appropriations of Descartes in both medicine and philosophy. But it also involved removing flaws in Stahl's animism. Next, in order to remain at the heart of eclecticism and to make his position acceptable to the dominant psychological spiritualism, he had to make Leibniz triumph over Stahl in the end, or show how Leibniz correctly imported lessons learned from Stahl into philosophy. Let us consider these two points.

Bouillier first explained that by appealing to the soul, insofar as it knows and reasons, when explaining the interplay of the organs, and in particular when establishing all the functions of life, Stahl himself remained too Cartesian. What had to be reused and valued in Leibniz was the theory that explained the unconscious nature of vital phenomena, that is to say, the theory of insensible perceptions, of blind and deaf thoughts.

References to the cases where consciousness breaks down, limit cases such as lethargy, apoplexy or even childhood, had been at the heart of the organicist argument. In his *Nouvelles études sur le spiritualisme*, for example, Pidoux shows how the theory of insensible perception had been tested at the *Académie impériale de médecine* where they experimented with induced anaesthesia (by inhalation of ether or chloroform) in order to observe what psychological phenomena occurred before, during, and after the administration of the anaesthetics. Their observations did not permit identifying the brain as the 'the organ of inner sense' (which would amount to endorsing materialism). They did however show that anaesthesia suppresses people's will and sense of unity.[22] Those experiments suggested that force should be resituated on the side of matter and, most importantly, they showed how the spiritualist re-appropriation of animism by Bouillier was a sleight of hand. For if some force or other did not subsist in those lethargic states, it could not originate from the soul, and consequently animism as conceived by Bouillier was proven wrong.

In relation to the question of consciousness, Bouillier's position thus appears as a veritable *tour de force*: he mobilizes the same argumentative

---

[22] On this point, see Pidoux, Nouvelles études sur le spiritualisme, 17–18.

framework as the one employed by the mechanist Cartesians in the physical register by holding that one cannot explain the infinite variations in the states of consciousness otherwise than by appealing to a domain of insensible perceptions governed by the same laws as the domain of sensible ones. Taking departure in the polemics against Locke's empiricism in the *Nouveaux essais sur l'entendement humain*, Bouillier thus reconceptualizes the Leibnizian minute perceptions *from the viewpoint of the physicians*, that is, Bouillier's adversaries, according to an argument *a fortiori*: 'The metaphysician who underestimates the reality of insensible perceptions is similar to the physician who, when studying material nature, does not take into account imperceptible bodies and insensible movements' (Bouillier, *Histoire de la philosophie cartésienne*, 199). This additional argument permits Bouillier to specify what he himself understands by 'animism': it is a doctrine that stresses a form of 'inner sense'. Leibniz can thus, at the same time, be habilitated and protected against all materialist or even pantheist misuse.

As such, Leibniz constituted the very last mediation needed by Bouillier in order to produce what amounted to a veritable *coup de theater*, although most of the preparatory work had already been done by Lemoine: Bouillier overturned the Cartesian model, but this time the dominant one. For it was precisely via this internal sense that Bouillier finally rediscovered a 'good' Descartes. By means of this good Descartes he completely renewed the analyses of the phenomena of passion he had proposed in the *Histoire de la philosophie cartésienne*, and even those proposed in chapter XII of the dissertation on the vital principle, dedicated to the 'transformation' of life in mechanism:

> [...] modern psychology has much too often forgotten the ancient theory of the inner senses, these senses that the organs do not seem to be exterior to, and that do not require that any external object be actually present. Descartes, however, and not going any further than to him, had noted them in ancient philosophy, and had provided a place for them in his theory of man. Besides the external senses, he allows for two internal senses that he defines in the following way: 'The first sense that I call interior comprises hunger, thirst and other natural appetites, and it is aroused in the soul by the nerves of the stomach, the gullet and all the other parts that serve the natural functions on account of which one has such appetites. The second comprises joy, sadness, love, anger and all the other passions, etc.' The reality of this internal sense which according to Descartes is aroused by the movements of the stomach, the gullet and all the other parts that serve the natural functions, has been recognized by a great number of physiologists who, in general, have concerned themselves more than the psychologists with the relations of the soul with the body.
>
> (Bouillier, *Histoire de la philosophie cartésienne*, 364–5)

When writing this, Bouillier pointed to a whole new field of study that his spiritualist friends had ignored: that of the soul's passions, those passions upon which the morality Cousin had wished for relied. The interpretation of Descartes that Bouillier finally endorsed – at the endpoint of a careful deconstruction of an institutionalized Descartes that had become so influential that no Cartesian alternative was available to critics such as Rochoux – is thus the Descartes we encounter at places in the *Traité de l'homme*, in the fifth part of the *Discours de la méthode*, at the end of the fourth part of the *Principes de la philosophie*, in the correspondence, with Elisabeth in particular, and, above all, in the treatise on the *Passions de l'âme*. It is a Descartes conceived via a new interpretation of Leibnizianism, saved from the official Cousinianism caricatural conception of its relation to 'animism' and 'Cartesianism'.

6. CONCLUSION

What does the example of Bouillier tell us about the reception of Leibniz in France in the mid-nineteenth century? It must first be underlined that this reception remains governed by the available texts and translations, and caught up in the reception of other authors through whom it is mediated (e.g. in the polemics surrounding Stahl). It is thus not Leibniz alone, or just Leibniz, who is received, but a Leibniz inserted into an intellectual constellation outside of which the various ways in which he is used remain perfectly incomprehensible. Second, in the context of the dialogue between spiritualism and the life sciences, we note, at the same time, an increasing interest and very strong distrust with regard to the monadology. In this context, the confrontation of the monadological theses with other theses in the Leibnizian system becomes indicative of the degree of autonomy that an author has with regard to an institutionalized thought within which monadology has no place.

More generally, Bouillier's intellectual itinerary teaches us that any reception of an author or a body of texts will be biased by institutional and mediations the effects of which show up in the way in which their history is told. Whether one chooses to mobilize the Leibniz of the monadology, pre-established harmony or minute perceptions, depends on the strategies that one adopts when entering one's own intellectual space – a space where such references already have several meanings and fulfil precise functions. In other words: interpretations of early modern authors are always refracted in the context of those who read them and they reflect the contemporary issues with which they are concerned, and this has consequences that can be identified and even interpreted in their texts.

Second, the references to an author do not acquire their meaning in and by themselves, but always in relation to references to other authors: Leibniz is interpreted in relation to interpretations of Descartes or Stahl – interpretations of Descartes and Stahl that are coloured by present-day concerns, that select, order, and recompose the doctrinal elements in those authors in ways that serve the dominant cause. Bouillier's work, in short, poses the question of the 'objectivity' and the 'autonomy' of the historian of philosophy.[23]

BIBLIOGRAPHY

Antoine-Mahut, Delphine. 'Cartésianisme dominant et cartésianismes subversifs. Le cas de l'"infirmier de Bicêtre Jean-André Rochoux'. *Corpus*, 68, 2015, forthcoming.

Antoine-Mahut, Delphine. 'La fabrique de l'histoire du cartésianisme néerlandais dans les histoires de la philosophie française du dix-neuvième siècle'. In *Les Pays-Bas aux XVIIe et XVIIIe siècles. Nouveaux Regards*, edited by C. Secretan and D. Antoine-Mahut, 107–24. Paris: Champion, 2015.

Antoine-Mahut, Delphine. 'Metaphysics, Psychology and Physics in the 19th Century French Histories of Philosophy'. In *Physics and Metaphysics. Issues and Directions in the Study of Cartesianism and Anti-Cartesianism*, edited by S. Roux and D. Antoine-Mahut. New York: Oxford University Press, 2016, forthcoming.

Biran, Maine de, Marie François Pierre Gonthier. *Examen des leçons de philosophie de M. Laromiguière*. Paris: Fournier, 1817.

Biran, Maine de, Marie François Pierre Gonthier. *Exposition de la doctrine philosophique de Leibniz*. Paris: L-G. Michaud, 1819.

Bouillier, Francisque. *Histoire et critique de la révolution cartésienne*. Lyon: L. Boitel, 1842.

Bouillier, Francisque. *Histoire de la philosophie cartésienne*. 2 tomes. Paris: Durand/Lyon, Brun et Cie, 1854.

Bouillier, Francisque. *De l'unité de l'âme pensante et du principe vital*. Paris: Auguste Durand/Lyon: Brun et Cie, 1858.

Bouillier, Francisque. *Du principe vital et de l'âme pensante, ou Examen des diverses doctrines médicales et psychologiques sur les rapports de l'âme et de la vie*. Paris: J-B. Baillère et Fils, Londres et Madrid, 1862.

Cotten, Jean-Pierre. 'Spinoza et Victor Cousin'. In *Spinoza au XXe siècle*, edited by A. Tosel, P.-F. Moreau and J. Salem, 231–42. Paris: Publications de la Sorbonne, 2005.

Cousin, Victor. *Oeuvres de Descartes*, en onze volumes. Paris: Pichon et Didier, 1824/1826.

Cousin, Victor. *Cours d'histoire de la philosophie moderne*, année 1815–1816. Paris: Ladrange, 1841.

Cousin, Victor. *Histoire générale de la philosophie depuis les temps les plus reculés jusqu'à la fin duXVIIIe siècle*. Paris: Didier et Cie, 1863.

---

[23]I am grateful to Mike Beaney, Jeremy Dunham, Mogens Laerke, Samuel Lézé, Pauline Phemister and the two anonymous reviewers for their relevant and stimulant comments on the manuscript. This paper was translated by Mogens Laerke.

Damiron, Jean-Philibert. *Essai d'histoire de la philosophie en France au dix-neuvième siècle*. Paris: F. Didot, 1828.
Damiron, Jean-Philibert. *Essai sur l'histoire de la philosophie en France au dix-septième siècle*. Paris: Hachette, 1846.
Foucher de Careil, Louis Alexandre. *Réfutation inédite de Spinoza par Leibniz. Précédée d'un mémoire par Louis Alexandre Foucher de Careil*. Paris E. Brière, rue Ste Anne 55, 1854.
Foucher de Careil, Louis Alexandre. *Leibniz, Descartes et Spinoza. Avec un Rapport par M. V. Cousin*. Paris: Ladrange, 1862.
Janet, Paul. 'Une nouvelle phase de la philosophie spiritualiste'. *Revue des deux Mondes* 108 (1873): 365–9.
Janicaud, Dominique. *Ravaisson et la métaphysique. Une généalogie du spiritualisme français*. Paris: J. Vrin, 1997.
Laprade, Richard de. *Animisme et vitalisme* (Mémoires de l'Académie de Lyon, classe des lettres, 1861).
Latreille, Camille. *Francisque Bouillier. Le dernier des cartésiens. Avec des lettres inédites de Victor Cousin*. Paris: Hachette et Cie, 1907.
Lemoine, Albert. *L'Animisme et le vitalisme de Stahl*. Paris: Germain-Baillière, 1864.
Leroux, Pierre. 'De la mutilation d'un écrit posthume de M. Jouffroy'. *Revue indépendante* (1er novembre 1842): 257–322.
Macherey, Pierre. 'Leroux dans la querelle sur le panthéisme'. *Cahiers de Fontenay* 36–8 (1985): 215–16.
Moreau, Pierre-François. 'Spinoza et Victor Cousin'. *Archivio di filosofia* 1 (1978) I: 327–31.
Moreau, Pierre-François. 'Spinozisme et matérialisme au XXe siècle'. *Raison présente* 51 (1979): 85–94.
Moreau, Pierre-François. 'Saisset lecteur de Spinoza'. *Recherches sur le XVIIe siècle* 4 (1980): 85–97.
Moreau, Pierre-François. 'Trois polémiques contre Victor Cousin'. *Revue de métaphysique et de morale* 4 (1983): 542–8.
Moreau, Pierre-François. 'Les enjeux de la publication en France des papiers de Leibniz sur Spinoza'. *Revue de métaphysique et de morale* 2 (1988): 215–32.
Nourrisson, Jean-Félix. *La Philosophie de Leibniz*. Paris: Hachette, 1860.
Pidoux, Claude. *Nouvelles études sur le spiritualisme. Le spiritualisme organique*. Paris: Asselin, 1869.
Rochoux, Jean-André. *Épicure opposé à Descartes*. Paris: Joubert, 1843.
Stahl, Georg Ernst. *Negotium otiosum, seu Skiamachia, adversus positiones aliquas fundamentales Theoriæ veræ medicæ a Viro quodam celeberrimo intentata, sed adversisarmis conversis enervata*. Halle: Litteris & Impensis Orphanotrophei, 1720.
Vermeren, Patrice. 'La philosophie au présent: le juif Spinoza (l'institution philosophique et la doctrine maudite du juif Spinoza)'. *Lignes* 12 (1990): 167–80.
Vermeren, Patrice. *Victor Cousin. Le Jeu de la philosophie et de l'État*. Paris: L'Harmattan, 1995.

ARTICLE

LEARNING FROM LEIBNIZ: WHITEHEAD (AND RUSSELL) ON MIND, MATTER AND MONADS[1]

Pierfrancesco Basile

Whitehead's system may be interpreted as a majestic attempt at recasting Leibniz's theory of monads in terms of sounder ontological categories. After a brief introductory section on the sources of Whitehead's knowledge of Leibniz's philosophy, the paper explains why Whitehead turned to Leibniz for metaphysical inspiration. Attention then shifts to Whitehead's understanding of the problems involved with Leibniz's theory of monads and his alternative explanation of monadic causation. Whitehead's endeavour to install windows in Leibniz's monads may not be entirely convincing, but there are philosophical gems scattered here and there in his analyses – true moments of insight that repay close examination.

1. INTRODUCTION

Like many of his contemporaries, Whitehead believed that Leibniz had provided a correct account of the ultimate principles of reality with his theory of monads.[2] But in the theory of pre-established harmony he saw the greatest stumbling block to its acceptance. In his metaphysical trilogy – *Science and the Modern World* (1925), *Process and Reality* (1929), and *Adventures of Ideas* (1933)[3] – he constructed a bewildering complex theory of reality and our place in it. In these books, he also offered an analysis of the strengths and weaknesses of Leibniz's theory of monads, together with a positive account of monadic interaction.

---

[1]Thanks to Pauline Phemister, Leemon B. McHenry, Jeremy Dunham, two anonymous referees and the Editor of this journal for comments on a previous version of this paper.
[2]Monadistic metaphysics are also advocated in Ward (*Realm of Ends or Pluralism and Theism*), Carr (*Theory of Monads, Cogitans Cogitata*). Even absolute idealists such as F. H. Bradley took Leibniz seriously, as his theory of monads can be viewed as a *pluralistic* alternative to their own brand of *monistic* idealism; cf. Basile ('Bradley's Metaphysics'), 189–208.
[3]For a full understanding of Whitehead's metaphysics, one should also consult some of his shorter and less ambitious books, especially *Religion in the Making* (1926), *Symbolism: Its Meaning and Effect* (1927), *The Function of Reason* (1929), and *Modes of Thought* (1938).

Whitehead was a man of impressive talents and erudition. He also had the good luck of working in Cambridge in close personal contact with exceptional colleagues and at a time when both science and philosophy were there flourishing.[4] But he was not trained either as a philosopher or a historian of philosophy. We know little or nothing about his private studies and ruminations. The appreciative tenor of many of his passages[5] suggests that he must have read some of Leibniz's writings quite carefully. To the best of my knowledge, however, all of his allusions to Leibniz in his printed works are to the *Monadology*.[6]

As far as scholarly knowledge of Leibniz's philosophy is concerned, Whitehead apparently owes it to two books. One is Louis Couturat's *La Logique de Leibniz* (1903), the other Bertrand Russell's *A Critical Exposition of the Philosophy of Leibniz* (1900). Both thinkers emphasize the significance of Leibniz's logic for a general understanding of his philosophy. Leibniz's theory of monads, Russell says, looks like a fairy tale unless one realizes that it follows from a few basic assumptions – the most important of which is the logical theory that all propositions are subject–predicate in form. As we shall see, Russell's interpretation profoundly shaped Whitehead's understanding of Leibniz's metaphysics.[7]

2. WHITEHEAD ON THE ECLIPSE OF SPECULATIVE REASON IN EARLY MODERN SCIENCE

Whitehead explains what he takes to be the contemporary relevance of Leibniz's metaphysics in *Science and the Modern World*. One of the factors that

---

[4]For biographical information, see Lowe (*Alfred North Whitehead*). A critical edition of Whitehead's writings is currently in preparation; work on this project is likely to lead to the discovery of new interesting material.

[5]Here is a notable example:

> Leibniz inherited two thousand years of thought. He really did inherit more of the varied thoughts of his predecessors than any man before or since. His interests ranged from divinity to political philosophy, and from political philosophy to physical science. These interests were backed by profound learning. There is a book to be written, and its title should be, *The Mind of Leibniz*.
>
> (MT 3)

[6]The word 'allusion' is used here because one would search in vain for a systematic discussion of Leibniz's thought in Whitehead's works; all Whitehead offers are occasional commentaries.

[7]References to Leibniz occur also in Russell's *The Principles of Mathematics* (1903), a book Whitehead had surely studied very carefully. Whitehead may have also read Robert Latta's *Leibniz: The Monadology and Other Philosophical Writings* (1898); the book contains a long introduction to Leibniz's thought as well as an anthology of texts, not just the *Monadology* and other metaphysical writings, but also the 'Introduction' to the *New Essays on the Human Understanding*, in which Leibniz steps forward as a phenomenologist *ante litteram*.

enabled the great scientists of the seventeenth and the eighteenth centuries to make their momentous discoveries, he contends, was their stubborn refusal to deal with basic speculative questions. In their physical investigations, they registered empirical regularities and accepted them as mere matters of fact. This procedure was justified pragmatically by the success of science in predicting the course of nature. But for a fully rational mind – one which takes seriously Leibniz's Principle of Sufficient Reason – this method cannot be satisfactory. Reason at its best craves understanding; it cannot acquiesce in the *factum brutum* of a mere acknowledged regularity.

In *Science and the Modern World*, Whitehead meditates upon the historical fate of this strong conception of rationality as the search for ultimate explanations. In this context, he discerns a deep meaning in the violent death of the Renaissance thinker Giordano Bruno. This occurred just before the rise of modern science:

> Giordano Bruno was the martyr: though the cause for which he suffered was not that of science, but that of free speculative imagination. His death in 1600 ushered in the first century of modern science in the strict sense of the term. In his execution there was an unconscious symbolism: for the subsequent tone of scientific thought has contained distrust of his type of general speculativeness.
> (SMW 1)

The eclipse of speculative reason in early modern thought is best exemplified in the way the question of the ultimate nature of things was approached. Modern science adopted as its ontological framework a materialistic atomism uncritically derived from antiquity. For the limited purposes of ordering and predicting phenomena, this ontology worked perfectly. But fundamental questions remained unanswered.

On a materialistic basis, for example, it is impossible to understand why there should be living, experiencing organisms. There is nothing in the idea of inert bits of matter that enables us to *see* why they should give rise to life and experience. The mutual relationships of material particles may be conceived as becoming increasingly complicated in the course of evolution. The problem of *understanding* the transition from the purely material aggregate to the living sentient organism still remains. 'A thoroughgoing evolutionary philosophy', Whitehead writes, 'is inconsistent with materialism. The aboriginal stuff, or material, from which a materialistic philosophy starts is incapable of evolution' (SMW 107). For Whitehead, you cannot be a materialist and believe yourselves to be rational – not if the word 'rational' is understood in its deep, speculative sense.

Whitehead's anti-materialist reasoning bears a strong similarity to Leibniz's Mill-Argument in the *Monadology*, an argument that will have hardly failed to attract Whitehead's attention although he does not discuss it explicitly. Contemplate the material machine as much as you will, enlarge it and step into it in your imagination. Still, you will not find anything there that *explains* why the material machine should *also* be thinking. In a letter to Lady Masham, commenting upon Locke's suggestion that God may endow matter with a power to think, Leibniz makes the point even more forcefully:

> It is true that the illustrious Locke maintained in his excellent *Essay* ... that God can give matter the power of thinking, because he can make everything we can conceive happen. But then matter would think only by a perpetual miracle, since there is nothing in matter in itself, that is, in extension and impenetrability, *from which thought could be deduced*, or upon which it could be based.
> (AG 290; my emphasis)

God could give matter the power of thinking; *matter* could never generate thought. Why? If there is nothing to matter but extension and impenetrability, the origination of mind out of matter would be a brute fact. If matter and mind happen to be thus conjoined, this connection would have to be established by an external agency (God) as well as *explicable* by reference to it:

> God, in the case of thinking matter, must not only *give* matter the capacity to think, but he must also *maintain* it continually by the same miracle, *since this capacity has no root*, unless God gives matter a new nature.
> (AG 290; my emphasis)

Leibniz and Whitehead are addressing what is nowadays called the 'hard' problem of consciousness. Their argument is a very powerful one. If consistently carried through it leads to the conclusion that, in one way or another, mind must be a fundamental feature of reality. As Whitehead has it, the principles of things cannot be 'vacuous actualities' (PR xiii), entities entirely devoid of mentality.

This is, in brief, how the Leibniz–Whitehead critique of materialism works:

(1) There must be an explanation of everything that is – of its very existence as well of its specific way of being.
(2) Mental events (such things as thoughts, memories, sensations, emotions) undoubtedly exist.
(3) There is obviously nothing in the concept of a material (merely extended) atom or in that of their combination into larger material aggregates that explains why mental events shall exist.
(4) Hence, materialism – the view that only material atoms truly are and anything else must be explained by reference to them – cannot be the last word about the nature of things.

Of course, such an argument is not immune from attacks. It seems natural to ask (1) why the Principle of Sufficient Reason should be accepted in the first place (which would lead to a complicated discussion of what has to be regarded as an 'explanation'). Eliminative materialists may want to reject (2), while believers in radical emergence would deny (3). But the reasoning surely has a strong *prima facie* plausibility and is as good as any to be found in philosophy.

Whitehead knew very well that mere philosophical reasoning would never make a successfully developing science deviate from its course. But he lived in a time of scientific revolution. Like many of his contemporaries – and because of his solid scientific education, probably much better than most – Whitehead realized that recent developments in biology (the theory of evolution) and physics (quantum physics, relativity theory) had destroyed any basis for believing in a theory of reality as simple as seventeenth-century materialism. This theory, he argued, is 'entirely unsuited to the scientific situation at which we have now arrived' (SMW 17). Philosophers, he concluded not without polemical undertones, have neither to indulge in historical studies nor provide detailed but sterile analyses of isolated technical problems; rather, they have to engage in the construction of a new metaphysical scheme (cf. PR xiv).

3. ON A FUNDAMENTAL TENSION IN LEIBNIZ'S METAPHYSICS

The one book in which Whitehead makes the utmost effort to be both complete and precise in the elaboration of his new ontology is *Process and Reality*. 'On the whole', Whitehead says here, 'this is the moral to be drawn from the *Monadology* of Leibniz. His monads are best conceived as generalizations of contemporary notions of mentality' (PR 19). It seems natural to read in this passage an allusion to an argument Leibniz develops in the three opening paragraphs of the *Monadology*.

> The Monad, which we shall discuss here, is nothing but a simple substance that enters into composite – simple, that is, without parts. And there must be simple substances, since there are composites; for the composite is nothing more than a collection, or *aggregate*, of parts. But where there are no parts, neither extension, nor shape, nor divisibility is possible. These monads are the true atoms of nature and, in brief, the elements of things.
> (§§1–3, AG 213)

Leibniz tells us here that the basic principles of reality (monads) cannot possibly be extended. Extended things are infinitely divisible, whereas the ultimate constituents of reality (substances) cannot have parts. The material (extended) atoms of modern science are useful abstractions, yet they cannot be regarded as metaphysically ultimate.

As Leibniz writes in a Letter to De Volder, 'the atoms of the Democriteans ... cannot, as commonly understood, be found in nature, nor are they anything but *incomplete thoughts of philosophers* who have not inquired sufficiently well into the nature of things' (AG 175; my emphasis). Whitehead agrees with this evaluation. On his understanding, science takes an *abstract* view of the nature of things. Scientific theories provide us with a description of the *structural* aspects of reality, but say nothing about the intrinsic nature of the things that support that structure. 'Science ignores what anything is in itself. Its entities are merely considered in respect to their extrinsic reality, that is to say, in respects to their aspects in other things' (SMW 153). To conceive of scientific concepts as adequate characterizations of things is to mistake the abstract (i.e., the 'incomplete') for the concrete – a mistake Whitehead refers to as 'the fallacy of misplaced concreteness'.

The argument against materialism based upon the Principle of Sufficient Reason that has been considered in the previous section is ambiguous as to its positive metaphysical implications. The view that purely material particles cannot account for the existence of mind shows that the concept of matter as sheer extension requires revision. By itself, however, the argument provides no clue as to the kind of correction that is needed. That line of reasoning is consistent both with the view (1) that the basic principles of reality have a mental as well as a physical side, and (2) the more radical view that such principles are purely mental.[8] The argument at the beginning of the *Monadology* would seem to have clearer metaphysical implications. In order to reach the basic principles of reality, we are now apparently told, we have to bypass the realm of the material altogether.

In this way, Leibniz may be read as committing himself to a form of *panpsychistic idealism* or, as it is perhaps better called, *mentalism* – the doctrine that monads are mind-like entities. On this view, our own mind becomes the paradigm of substantiality.

> If we wish to call *soul* everything that has *perceptions* and *appetites* ... then all simple substances or created monads can be called souls.
>
> (§19, AG 215)

Leibniz's descriptions of the monads are brief but suggestive. Each monad is a unique synthesis of perceptions (§78; AG 223). Moreover, their inner life is on-going activity (§11; AG 214), as each monad is driven by an inner desire (*appetition*) for novel experiences (§15; AG 215).[9] Whitehead applauds at this understanding of the monad as an experiential process-unit. 'Each monadic creature is a mode of process of "feeling" the world,

---

[8]These two options clearly parallel the debate in Leibniz studies over whether Leibniz is a corporeal substance realist or a pure idealist. This complex debate cannot be adjudicated here. Whitehead himself does not seem to recognize that the concept of the monad can be interpreted in two distinct ways.

[9]As Leibniz explains to De Volder, 'the true notions of things are completely turned on their heads by that new philosophy which forms substances from what is only material and passive' (AG 174).

of housing the world in one unity of complex feeling, in every way determinate' (PR 80). But of his own metaphysics, he also says:

> This is a theory of monads; but it differs from Leibniz's in that his monads change. In the organic theory, they merely become.
>
> (PR 80)

For the time being, it is important to notice the point of distinguishing sharply between the concept of *change* and the concept of *becoming*. This highlights what Whitehead takes to be a fundamental tension in Leibniz's metaphysics. On the one hand, Leibniz's monads are mind-like currents of experience. On the other, Whitehead thinks, they are conceived by Leibniz in an Aristotelian fashion as substrata to which qualities inhere. On this view monads 'change' – that is, each monad is a permanent substratum that loses old properties and acquires new ones.

Whitehead is unsatisfied with this Aristotelian account of the monads' ontological structure for several reasons. How is it possible to conceive of *energetic* monads in terms of the *static* notion of an underlying bearer? The notion of substance as a substratum clashes with Leibniz's other fundamental idea that a substance is *essentially* a being capable of action. And how could categories such as those of 'thing' and 'properties' be of any use in describing immaterial beings? These categories had been developed by primitive men in primitive times to deal with ordinary material objects such as rocks, trees and mountains. These concepts are of great practical utility, but they were not meant to have any deep ontological significance.

But the main obstacle that prevents understanding monads as substrata, Whitehead argues, is their experiential nature. The concept of the monad as embodying a point of view is a thoroughly relational concept. We *experience* the world by grasping and incorporating aspects of it into the unity of a new perspective. This process must involve for Whitehead some sort of actual relationship between the experiencing subject and the experienced object. But there is no place for relations within a metaphysical scheme that acknowledges only the reality of properties and their underlying bearers.

Within such a scheme, Whitehead contends, reality collapses into a plurality of mutually isolated substances:

> The doctrine of the individual independence of real facts is derived from the notion that the subject-predicate form of statement conveys a truth which is metaphysically ultimate.
>
> (PR 137)

As Whitehead sees things Leibniz has grasped the truth about the ultimate constituents of reality, but lacks the conceptuality in which to express it. Worse than this, the concepts he eventually adopts – the notions of a substratum and its properties – embody a metaphysical point of view which is the exact opposite of the one he wishes to express. These Aristotelian notions privilege permanence over flux, inertness over activity and, more importantly for the purposes of the present paper, mutual separateness over relatedness. Only with the 'deposition of substance-quality ontology', he contends, 'we can reject the notion of individual substances, each with its own private world of qualities and sensations' (PR 160).

According to Whitehead, it was precisely in order to soften this tension at the heart of his metaphysics that Leibniz developed the doctrine of pre-established harmony, according to which the monads' perceptions have been synchronized by God. This theory enables Leibniz to hold that monads are experientially connected, while at the same time denying that they are causally related:

> [Leibniz] had ... on his hands two distinct points of view. One was that the final real entity is an organizing activity, fusing ingredients into a unity, so that this unity is the reality. The other point of view is that the final real entities are substances supporting qualities. The first point of view depends upon the acceptance of ... relations binding together all reality. The latter is inconsistent with the reality of such relations. To combine these two distinct points of view, his monads were therefore windowless; and their passions merely mirrored the universe by the divine arrangement of a pre-established harmony.
> (SMW 155)

In Whitehead's eyes, the theory of pre-established harmony is an artificial device the only purpose of which is to conceal a manifest contradiction in the system.[10]

Because of the strong connection between Leibniz's theory of pre-established harmony and his belief in God's existence, some readers may suspect at this point that Whitehead's rejection of that theory may be motivated by a typically modern dislike of God as an explanatory principle. As a matter of fact, Whitehead believes in God, although his God is not that of traditional Christianity; there must be, he argues in *Process and Reality*, a radical new way of conceiving of the Deity and his relation to the world after Hume's devastating critique in the *Dialogues Concerning Natural Religion* (PR 343). As the following two sections will further make clear, the

---

[10] It could be argued that this contradiction will necessarily reappear within Whitehead's system as well, only in a slightly different form. Since the substance-predicate mode of thought is reflected in the grammatical structure of our language, Whitehead has the problem of formulating his metaphysical vision in terms of a linguistic medium that embodies those very metaphysical commitments he explicitly rejects. Whitehead is fully aware of this difficulty; eventually, he contends, the metaphysician will have to 'redesign' ordinary language, so as to make it adequate to his or her own purposes (cf. PR 11). But it should be noted that this problem is not unique to Whitehead's philosophy; rather, it will arise within any system of revisionary metaphysics. On this point, see the brief but illuminating remarks in Kraus (*Metaphysics of Experience*), 1–8, as well as the insightful Simons ('Metaphysical Systematics').

reasons for his rejection of Leibniz's pre-established harmony have nothing to do with a prejudiced hostility towards theology, but are quintessentially philosophical.[11]

4. RUSSELL AND WHITEHEAD ON THE METAPHYSICS OF SUBSTANCE

In his understanding of Leibniz's monads as substrata of change Whitehead is deeply influenced by Russell's *A Critical Exposition of the Philosophy of Leibniz*. Russell denounces here the metaphysical inadequacy of substance-property ontology, arguing that only two main types of metaphysical systems can be constructed on this basis – either Leibniz's theory of a plurality of independent monads or Spinoza's theory of a single encompassing Reality. Since both theories are plainly false, Russell goes on to argue, the metaphysician has to reject the subject–predicate mode of thought and devise radically new ontological categories:

> Spinoza ... had shown that the actual world could not be explained by means of one substance; Leibniz showed that it could not be explained by means of many substances. It became necessary, therefore, to base metaphysics on a notion other than that of substance – a task not yet accomplished.
> (CEPL 126)

Whitehead repeats the argument almost *verbatim* in a number of places, for example in *The Concept of Nature*:

> Some schools of philosophy, under the influence of the Aristotelian logic and the Aristotelian philosophy, endeavour to get on without admitting any relations at all except that of substance and attribute. Namely all apparent relations are to be resolvable into the concurrent existence of substances with contrasted attributes. It is fairly obvious that Leibnizian monadology is the necessary outcome of any such philosophy. If you dislike pluralism, there will be only one monad.
> (CN 150)

Stated in its explicit form, Russell's critique of the substance-predicate mode of thought runs as follows:

(1) Relations cannot be analysed in terms of the concepts of substance and property. (As a way of brief illustration, consider the statement 'A is higher than B.' It would seem natural to analyse it in terms of the

[11]For an accessible introduction to Whitehead's novel conception of God, as well as for those main aspects of his thought not discussed in this paper, see Sprigge (*The God of Metaphysics*), 409–72.

conjunction 'A is of height X' and 'A is of height Y.' But Russell holds that this analysis is incomplete unless one also specifies what kind of relation holds between the two heights X and Y.)
(2) Hence, if the ultimate constituents of things are bearers of properties, relations are to be condemned as unreal.
(3) On this basis, the world can be conceived either as a single Substance (in the way of Spinoza) or as a plurality of unrelated substances (in the way of Leibniz).
(4) Since both theories (Spinoza's Monism and Leibniz's Pluralism) are patently absurd, the doctrine that only substrata and their properties are real must be rejected as false.

In his discussion of this argument, Russell focuses upon (1) and (2). Russell is especially plagued by the question whether it is true to say that Leibniz denied that relations are real, for Leibniz does not say that they are 'unreal', but that they are 'ideal'. According to Russell, Leibniz saw that relations cannot be reduced to properties. Having gone so far, he should have acknowledged relations as among the basic constituents of reality. His obstinate belief in the subject–predicate logic prevented him to do so. Torn between his own insights and his traditional Aristotelian background, he looked for a compromise solution. Hence, he ended up arguing that relations are 'ideal' – that is, 'half-real' or 'semi-real'. For Russell this is only a subterfuge of no philosophical significance, one that is not worthy of a great logician.[12]

With respect to (3), Russell observes that Leibniz's Monadism and Spinoza's Monism are not logically on a par, as only Spinoza's is a coherent form of substance-metaphysics. The reason for this is that Leibniz believes his monads to be components of the very same universe. But there is no way to make sense of the idea of their unity within one world without introducing the notion of their being related. In order to achieve unity without recourse to relations, Leibniz would have to reduce *all* monads to adjectives of a *single* Substance; in this way, he would have become himself a Monist.

In due time, this early criticism of the traditional Aristotelian theory will lead to the development of a relational logic in *Principia Mathematica* (1910–13). The implications of Russell's early critique are developed further in *Process and Reality,* where Whitehead explains that the new relational logic will have to be accompanied by a new, relational metaphysics. In this book, he writes, '"relatedness" is dominant over "quality"' (PR xiii).

---

[12]This aspect of Russell's critical interpretation of Leibniz's philosophy has received much attention by commentators. See, for example, Parkinson (*Logic and Reality in Leibniz's Metaphysics*), Ishiguro ('Leibniz's Theory of the Ideality of Relations'), Mates (*Philosophy of Leibniz*), Rescher ('Leibniz on Intermonadic Relations'), and Mugnai (*Leibniz's Theory of Relations*).

Surprisingly, Russell says little or nothing to justify (4), the thesis that Monism and Monadism are patently false. And yet this claim is crucial if the above argument is to function as a *reductio*. Whitehead sees that this point stands in need of explanation. What makes the metaphysics of Spinoza and Leibniz utterly incredible, he says, is that 'either alternative stamps experience with a certain air of illusoriness' (PR 190).

This charge has to be understood with reference to Whitehead's theory of experience. According to him, we do constantly perceive that there are *many things* around us (which rules out Monism) as well as that *they act upon us* in a multitude of ways (which rules out causally independent monads). In conversation with one of his Harvard colleagues, Whitehead made the point thus: '*Being tackled at Rugby, there is the Real*. Nobody who hasn't been knocked down has the slightest notion of what the Real is.'[13] In having such a violent experience we feel absolutely certain – with a vividness we cannot possibly resist – that there are independent causal powers taking hold of us.

The view that we have a direct apprehension of external causal powers is the view of common sense. Philosophically, it is a bold one. After Hume, philosophers have been accustomed to think of our experience of causal processes as a matter of experiencing a succession of distinct events. This may be the case when one observes two objects acting upon one another, as in Hume's favourite example of the two striking billiard balls. But things change radically when *we* ourselves are involved in the causal transaction. In these cases, we are able to observe the workings of causation from 'within'. This happens, for example, when a light makes a man blink: 'The man will explain his experience by saying, "The flash made me blink"; and if his statement be doubted, he will reply "I know it, because I felt it"' (PR 175).

Of course, we cannot expect to have a clear and distinct sensory image of such powers in the same way in which we have a clear and distinct sensation of red when we look at a red object. But for Whitehead this only shows that Hume's analysis of ordinary human experience in terms of sensory impressions is unduly narrow. The kind of experience we have when we fall on the ground or are violently beaten with some hard object can hardly be described as an apprehension of distinct sense-impressions. A distinction must therefore be drawn between 'perception in the mode of presentational immediacy' (which covers awareness of sense-data) and 'perception in the mode of causal efficacy' (which covers awareness of causal force).

This distinction seems unobjectionable. There is really a dynamic dimension to our experience that traditional empiricist theories fail to do justice to.[14] Experience is not just a matter of *looking* at things, but also of *acting* and *being acted upon* – we are not mere spectators, but *doers* as well. Still, appealing

---

[13] Hocking (1963), 15.

[14] Leibniz does, of course, admit as much in his insistence that the nature of substance consist in force and his doctrine of appetition; his monads have a dynamic inner experiential life.

to experiences of causal efficacy as a way of refuting Leibniz's doctrine of the causal insulation of the monads may appear argumentatively weak. Can we refute the doctrine of pre-established harmony in the way Dr Johnson thought he could refute Berkeley's immaterialism – just by kicking a stone? The answer to this question is, obviously, 'no'. Leibniz could easily grant that our experiences suggest causal efficacy, while at the same time denying that they bear witness to the existence of real causal connections.[15]

It must be admitted therefore that there is nothing in Whitehead's analysis of experience of the nature of a proof. The best way to read his account is to take it as an earnest invitation to consider one's experiences more closely and decide for oneself which interpretation is the more adequate.[16] Looking at his theory in this way, it becomes hard to deny that his account is superior to traditional empiricist ones. Moreover, it becomes plain that he is also raising a very serious question. Given that ordinary experience inevitably *suggests* a constant apprehension of causal agencies, how could we ever *believe* Leibniz's doctrine of the unreality of causal interaction? In order to believe this doctrine, we would have to silence our most natural epistemic instincts, which is surely not an easy thing to do.

5. LEIBNIZ AND WHITEHEAD ON THE NATURE OF CAUSATION

The Russell–Whitehead critique of Leibniz stands or falls with the claim that he conceived of his monads as bearers of properties/substrata of change. Russell provides little conclusive evidence in its support. Logical considerations are important in the *Discourse on Metaphysics*, but almost irrelevant in the *Monadology* and the *Principles of Nature and Grace*. Russell even admits that in the correspondence with De Volder Leibniz seems on the verge of adopting a process-metaphysics of events, as the monad almost dissolves into the mere series of its perceptual states.[17]

Whitehead seems to be unaware of this when he criticizes Leibniz's doctrine of the causal independency of the monads. In *Modes of Thought*, for example, he remarks:

---

[15] Analogously, Spinoza could retort that he is not denying the existence of many active beings, but solely arguing that such beings are to be regarded as 'modes', not as 'substances'. At any rate, Spinoza surely did not conceive of his one Substance as a bearer of properties/substratum of change, but rather in dynamic terms; cf. Basile (*Russell on Spinoza's Substance Monism*).

[16] An open and undogmatic mind, Whitehead suggests as much when he says: 'It is impossible to scrutinize too carefully the character to be assigned to the datum in the act of experience' (PR 157).

[17] In a letter to De Volder Leibniz speaks of the monad as 'something analogous to the soul, whose nature consists in a certain eternal law of the same series of change' (AG 173). For a recent discussion of Leibniz's account of the relationship holding between a monad and the law determining the series of its changes, see Whipple ('Structure of Leibnizian Simple Substances').

> The mere notion of transferring a quality is entirely unintelligible. Suppose that two occurrences may in fact be detached so that one of them is comprehensible without reference to the other. Then all notion of causation between them, or conditioning, becomes unintelligible.
>
> (MT 164)

This is an allusion to that crucial passage in the *Monadology* in which Leibniz explains why monads cannot be affected from without:

> The monads have no windows through which something can enter or leave. Accidents cannot be detached, nor can they go about outside of substances, as the sensible species of the Scholastics once did. Thus, neither substance nor accident can enter a monad from without.
>
> (§7; AG 213–14)

The passage may be read as providing some evidence in support of the thesis that monads are for Leibniz bearers of properties as well as that there is a close connection between this Aristotelian conception and the denial of direct causal interaction. Leibniz is here concerned with that type of causal fact in which a substance A acts upon an already existing substance B, so as to produce an alteration in B. Leibniz denies that monads can be related in this way by means of the following argument:

(1) Causation (in the sense of a substance's ability to produce an alteration in another substance's state) involves the transference of some one element from one substance A to another substance B.
(2) On an Aristotelian substance-predicate ontology, the transferred element must be either the active substance or one of its properties.
(3) Substances cannot become parts of other substances; for substances are simple, not complex. Hence, the transferred element will have to be a property.
(4) During the transaction from substance A to substance B, there will be a moment in which the transferred property does not belong to either A or B. This is impossible, since properties require a substratum in which to inhere.
(5) Hence, on an Aristotelian substance-predicate ontology, direct causal interaction between monads is impossible, as neither substances nor properties can be exchanged.

To someone convinced that causal relations are real, this reasoning will look like a further argument against the thesis (involved with premise (2)) that substances are substrata, which is precisely how Whitehead reads it.

What is noteworthy is that Whitehead does not interpret this argument, as he would be entitled to do from a logical point of view, as a *reductio* of premise (1) – the influx-model of causation that Leibniz derives from the late Scholastics. As a matter of fact, Whiteheads shares with Leibniz the notion that causation involves some kind of flowing in, a transference of elements from the causally active substance to the affected one. Trying to explain his rejection of direct monadic interaction to De Volder, Leibniz says: 'Properly speaking, I don't admit the action of substances on one another, since there appears to be no way for one monad to flow into another' (AG 176). The challenge of providing an alternative to the theory of pre-established harmony now becomes for Whitehead the question of how to make sense of monadic influx in categories different from those of substance and property.[18]

6. WHITEHEAD'S INTERACTING MONADS: THE CONCEPT OF *PREHENSION*

What remains of the monad once the concept of substratum has been rejected? Radically new ontological categories are needed and one main way to discover them is by examining the monad as this is concretely given. This is possible in a philosophy such as Whitehead's because monads are mind-like entities. Our own inner life shows what the fundamental constituents of reality are like.[19]

---

[18]Whitehead also charges Leibniz with the following contradiction:

> ... no reason can be given why the supreme Monad, God, is exempted from the common fate of isolation. Monads, according to this doctrine [i.e., pre-established harmony] are windowless for each other. Why have they windows towards God, and why has God windows towards them?
>
> (AI 134)

Whitehead would seem here to be guilty of some confusion. On the one hand, Leibniz's monads have no power to affect God; it is simply not true that God has 'windows toward them'. On the other hand, in the *Monadology*, Leibniz explicitly ascribes to God the power to create monads, to preserve them in existence and to destroy them. But these kinds of power are very different from the sort of causal power the possession of which Leibniz denies to his monads. The influx-model of causation provides an ontological analysis of a substance's power to *alter* the nature of another substance; the divine powers are of a quite different order.

[19]To the best of my knowledge, problems concerning the reliability of introspection are never discussed by Whitehead. In *Adventures of Ideas*, he raises the question 'how the structure of experience is directly observed' (AI 177), but fails to address it. Whitehead says that as a matter of fact he first arrived at his basic ontological concepts by way of his study of physics rather than of human psychology:

> It is equally possible to arrive at this organic conception of the world [i.e., Whitehead's new metaphysics] if we start from the fundamental notions of modern physics ... In fact

In consonance with James's account of the self in the *Principles of Psychology* (cf. SMW 143 ff.), Whitehead contends that our conscious life comes as a series of 'pulses' or 'throbs' of experience. The stream of consciousness is broken in a plurality of successive quanta of feeling, each of which lasts for a brief period of time. Such total moments of experience Whitehead calls 'actual occasions' (or, alternatively, 'actual entities') and offers them as substitutes for Leibniz's monads (cf. AI 177).

The new ontology Whitehead settles for is one of serially interconnected experiential events:

> The soul is nothing else than the succession of my occasions of experience, extending from birth to the present moment.
>
> (MT 163)

Needless to say, this will not suffice as an account of personal identity. A distinction needs to be drawn between the enduring self we ordinarily refer to as our 'I' and the fleeting momentary selves (the actual occasions) that constitute our experiential life; if one renounces the notion of substance as an enduring *substratum*, then the emergence of the former out of the latter stands in need of an explanation. In order to understand Whitehead's view as to the nature of causation, however, there is no need to enter into a discussion of this difficult problem.[20] The only relevant question concerns the nature of the relation holding between the momentary occasions.

How are such successive occasions related? Consider our experiences when we are reading a book in a silent room and a thunder suddenly breaks in. What we experience then is not merely the sound of the thunder, but the *breaking* of the previous silence through it. This would not be possible if the silence of a moment ago were not still included within the novel moment in which the breaking of the thunder is experienced. Or consider our hearing of a musical melody. We could not experience any music as a flowing piece if the notes just past were not also included in our present awareness.

These examples are meant to show that there is a kind of natural 'fusing' within our conscious stream of its successive experiential occasions. The kind of phenomenon Whitehead is drawing our attention to here is the one Husserl calls 'retention', an apt term that will be used in what follows. Whitehead goes beyond Husserl in that he provides a metaphysical as

> by reason of my own studies in mathematics and mathematical physics, I did ... arrive at my convictions in this way.
>
> (SMW 152)

It is what may be called the 'phenomenological' approach to metaphysics, however, that is particularly at evidence in Whitehead's writings.

[20] For a recent discussion, see Mingarelli ('Is Personal Identity Something That Does Not Matter?').

opposed to a mere phenomenological account of this phenomenon. Retention is for him the way in which an immediately past occasion of experience leaves a mark upon – and thereby conditions – a novel moment of experience. In other words, Whitehead suggests that we interpret retention as a form of mental causation – as an 'influx' of aspects of the past moment of experience within the novel one.

In order to understand Whitehead's view correctly it is important to realize that the term 'influx' is not used by him in a metaphorical sense. The elements retained in the novel occasion are actual components of the previous occasion of mentality, not mere *representation*s of them:

> The present moment is constituted by the influx of *the other* into that self-identity which is the continued life of the immediate past within the immediacy of the present.
>
> (AI 181)

Whitehead coined the term 'prehension' to designate this incorporation of aspects of our past experiences into the present one. The concept is introduced by way of a comparison with Leibniz:

> [Leibniz] employed the terms 'perception' and 'apperception' for the lower and higher ways in which one monad can take account of another, namely for ways of awareness. But these terms ... *are all entangled with the notion of representative perception which I reject* ... Accordingly, on the Leibnizian model, I use the term 'prehension' for the general way in which the occasion of experience can include, as part of its own essence, any other entity.
>
> (AI 234; my emphasis)

This is an intricate yet fundamental passage. In the first place, Whitehead emphasizes that retained contents are not duplicates (*representative perceptions*) of previous experiences, but the original experiences themselves. Secondly, he explains that the relation between two successive occasions within our stream of consciousness has a double nature. Viewed from the standpoint of the past occasion, the experiential influx involved in the phenomenon of retention is a basic form of causation. Viewed from the perspective of the novel occasion, that influx is to be regarded as an elementary form of perception. This goes at the very heart of Whitehead's critique of Leibniz, for Whitehead is putting together again in his concept of prehension what Leibniz wants to keep distinct with his theory of pre-established harmony: the capacity to perceive on the one hand, that of being directly acted upon on the other. Thirdly, Whitehead draws a comparison with Leibniz's concept of 'perception'. This is a way of emphasizing the fact that the type of experience he calls 'prehension' is a primitive, low-level one. However, this is not to say that this fundamental sort of experience lacks a sense of subjective enjoyment; it solely means that it can, yet need not be made, the object of a higher order thought or experience (Leibniz's 'apperception').

Whitehead's account of causation in terms of the double notion of influx/ prehension is a startling one. One first major problem with it is that the kind of mental determination envisaged by Whitehead seems insufficient to account for causal relations between occasions pertaining to other individuals' experiential streams. Whitehead argues, on the contrary, that his account can be generalized to *all* existing occasions. Stated in Leibnizian terminology, there is no fundamental distinction between 'immanent' and 'transeunt' causation – that is, causation understood as a relation between the inner occurrences of a single monad's life and causation understood as a relation between numerically distinct monads. The principles I discover in myself have universal validity:

> ... in so far as we apply notions of causation to the understanding of events in nature, we must conceive these events under the general notions which apply to occasions of experience. For we can only understand causation in terms of our observation of these occasions.
>
> (AI 184)

As a matter of fact, Whitehead thinks that we constantly retain contents originally pertaining to experiential streams other than our own, especially of the occasions of experience that constitute our body. An intimate relation holds between them and our mind, as our present experience contains those very experiences that were just felt by the several parts of the body. To feel a pain in one hand is on this view to experience (with some slight temporal delay) the very same pain that was felt by the hand. It is not that we merely localize the pain in a certain region of the body; the pain was originally felt *by the hand*, and only secondarily by *us*.

The human body, Whitehead remarks,

> ... is a set of occasions miraculously coordinated so as to pour its inheritance into various regions within the brain. There is thus every reason to believe that our sense of union with the body *has the same original as our sense of unity with our immediate past of personal experience*.
>
> (AI 189; my emphasis)

On this understanding, the body is constituted by myriads of interconnected centres of feeling, each of which (but especially those constituting the brain) stands in a privileged causal contact with that ruling centre which is our mind. This is meant to be a more realistic version of Leibniz's theory of the mind as the body's dominant monad,[21] since it vindicates the incoercible sense we all share that our mind somehow mingles with our body.

---

[21] In the case of human beings, Whitehead does in fact speak of 'dominant centres of feeling' (MT 22). Differences between different types of living organisms are explained by reference to the concept of a dominant centre: 'A vegetable is a democracy; an animal is dominated by

Still, on Whitehead's theory of causation there is no reason in principle why a person should not be able to retain elements *from other person's* minds, which sounds indeed little plausible. Not everyone will agree with this sceptical evaluation. The fact that Whitehead's theory accounts for the possibility of an interchange between occasions constituting different personal streams may be viewed by some as a promising matrix for understanding parapsychological phenomena or religious experiences.[22] Whitehead seems willing to admit telepathy as a real phenomenon (cf. PR 308). And at one point he denies that knowledge of other minds is possible only by way of interpretation of our immediate sensory perceptions. 'The claim that cognition of alien mentalities must necessarily be by means of indirect inferences from aspects of shape and of sense-objects is wholly unwarranted' (SMW 150).

Another objection to Whitehead's account is that it seems faulty even as an explanation of immanent causation. At one point, Whitehead clarifies the nature of that link thus:

> All relatedness [between occasions] ... is wholly concerned with the appropriation of the dead by the living – that is, with 'objective immortality' whereby what is divested of its own living immediacy becomes a real component in other living immediacies of becoming.
>
> (PR xiii–xiv)

As a way of illustration, consider the previous example of our hearing a piece of music. When we hear a melody, the notes just heard keep resonating in the present moment, yet they have lost their character of *presentness* (their 'living immediacy') and have acquired a character of *pastness* ('objective immortality'). The notes are apprehended as belonging to the past – as feeble versions or echoes of the originals. This is correct from a purely phenomenological point of view, yet Whitehead also provides a metaphysical interpretation of the phenomenon of retention as involving a real entrance of the past into the present. But how could the retained experience and the experience originally felt be numerically the same, if they come with a different qualitative feel? This is impossible. One cannot divorce the appearance of an experience from what that experience is in the way required by Whitehead's account; an experience, considered as such, simply *is* what it feels like.[23]

---

one, or more centres of experience' (MT 24). Inanimate objects such as chairs and rocks lack any dominant centre.

[22]This view is advocated in Griffin (*Unsnarling the World-Knot*, 206–7).

[23]Whitehead's entire critique of materialism as explained in Section 2 does indeed presuppose this phenomenological (and commonsensical) notion of experience. The following passage is long, but deserves to be quoted at full length:

7. CONCLUSION

In Whitehead's interpretation, Leibniz has grasped the truth about the ultimate principles of reality, yet he develops his theory of monads in a way that makes it both internally incoherent and empirically inadequate. While the first allegation is doubtful, the second is justified. Whitehead's bold and original attempt to ameliorate Leibniz's theory of monads by endowing them with windows, however, is not satisfactory either. Contrary to what he believes, it may not be so easy to dislodge the doctrine of pre-established harmony from Leibniz's metaphysics.

But is a Leibnizian theory of reality worthy of consideration in the contemporary philosophical climate? This question must be answered positively. Philosophers of mind are still struggling to find a place for consciousness within the physical world. The fact that human subjectivity resists incorporation within a materialist framework suggests that we lack any proper understanding of the nature of matter, rather than of mind. As Whitehead rightly noticed, Leibniz's critique of the metaphysics of early modern science is a simple yet majestic lesson in speculative thinking – one that we should take seriously even today.[24] All this must be understood with a grain of salt, of course, for we cannot conclude that Leibniz's metaphysics is true as its stands. Upon all those in search of an alternative to

---

> The notion of life implies a certain absoluteness of self-enjoyment. This must mean a certain immediate individuality, which is a complex process of appropriating into a unity of existence the many data presented as relevant by the physical process of nature. Life implies the absolute, individual self-enjoyment arising out of this process of appropriation. I have, in my recent writings, used the word *prehension* to express this process of appropriation. Also I have termed each individual act of immediate self-enjoyment an *occasion of experience*.
>
> (MT 150)

This makes it clear that for Whitehead an experience cannot be divorced from its qualitative aspect – that is, to experience is, literally, to 'enjoy'. He also provides the following refutation of behaviourist accounts of human experience: 'A consistent behaviorist cannot feel it important to refute my statements. He can only behave' (MT 23).

[24]Thomas Nagel has recently argued for the necessity to rethink our concept of nature in a way strongly reminiscent of both Leibniz and Whitehead. He contends not solely that 'mind is not just an afterthought or an accident or an add-on, but a basic aspect of nature' (Nagel, *Mind and Cosmos*, 16); he even explicitly justifies this statement by reference to the Principle of Sufficient Reason, holding in the same context that 'pure empiricism is not enough' (17). In Britain, Leibnizian and Whiteheadian themes have been insightfully discussed by Sprigge (*Vindication of Absolute Idealism*, *Importance of Subjectivity*). They have forcefully reappeared in Strawson ('Realistic Monism' and *Selves: An Essay in Revisionary Metaphysics*). In the USA, such views have been kept alive throughout the entire twentieth century by a relatively small yet vigorous school of process thinkers. Besides the already mentioned Griffin (*Unsnarling the World-Knot*), Hartshorne (*Creative Synthesis and Philosophic Method*, 'Physics and Psychics') is a philosopher particularly worthy of mention.

materialism, however, the terse paragraphs of Leibniz's *Monadology* will hardly cease to exert a deep, thought-provoking fascination.

ABBREVIATIONS

AG = Leibniz, G. W. *Philosophical Essays*, edited and translated by R. Ariew and D. Garber (Indianapolis, IN: Hackett, 1989).
CEPL = Russell, B. *A Critical Exposition of the Philosophy of Leibniz: With an Appendix of Leading Passages* (London: George Allen and Unwin, 1900).
CN = Whitehead, A. N. *Concept of Nature* [1920] (Cambridge: Cambridge University Press, 2000).
SMW = Whitehead, A. N. *Science and the Modern World* [1925] (New York: The Free Press, 1967).
PR = Whitehead, A. N. *Process and Reality: An Essay in Cosmology* [1929], corrected edition by D. R. Griffin and D. W. Sherburne (New York: The Free Press, 1978).
AI = Whitehead, A. N. *Adventures of Ideas* [1933] (New York: The Free Press, 1967).
MT = Whitehead, A. N. *Modes of Thought* [1938] (New York: The Free Press, 1969).

BIBLIOGRAPHY

Basile, P. 'Russell on Spinoza's Substance Monism'. *Metaphysica: International Journal for Ontology and Metaphysics* 13, no. 1 (2012): 27–41.
Basile, P. 'Bradley's Metaphysics'. In *The Oxford Handbook of Nineteenth-Century British Philosophy*, edited by William Mander, 189–208. Oxford: Oxford University Press, 2014.
Carr, H. W. *A Theory of Monads: Outlines of a Philosophy of the Principle of Relativity*. London: Macmillan, 1922.
Carr, H. W. *Cogitans Cogitata*. Los Angeles: The Favill Press, 1930.
Griffin, D. R. *Unsnarling the World-Knot: Consciousness, Freedom and the Mind-Body Problem*. Berkeley: University of California Press, 1998.
Hartshorne, C. *Creative Synthesis and Philosophic Method*. London: SCM Press, 1970.

Hartshorne, C. 'Physics and Psychics: The Place of Mind in Nature'. In *Mind and Nature*, edited by J. Cobb and D. R. Griffin, 89–96. Washington, DC: University Press of America, 1977.

Hocking, W. E. 'Whitehead as I Knew Him'. In *Alfred North Whitehead: Essays on His Philosophy*, edited by G. K. Kline, 1–17. Englewood Cliffs, NJ: Prentice Hall, 1963.

Ishiguro, H. 'Leibniz's Theory of the Ideality of Relations'. In *Leibniz: A Collection of Critical Essays*, edited by H. Frankfurt, 191–224. New York: Anchor Books, 1972.

Kraus, E. M. *The Metaphysics of Experience: A Companion to Whitehead's "Process and Reality"*. New York: Fordham University Press, 1988.

Leibniz, G. W. *The Monadology and Other Philosophical Writings*, edited by R. Latta. Oxford: Clarendon Press, 1898.

Lowe, V. *Alfred North Whitehead: The Man and His Work*. 2 vols. Baltimore, MD: Johns Hopkins University Press, 1985 and 1990.

Mates, B. *The Philosophy of Leibniz: Metaphysics and Language*. Oxford: Oxford University Press, 1986.

Mingarelli, E. 'Is Personal Identity Something That Does Not Matter? An Inquiry into Derek Parfit and Alfred N. Whitehead'. *Process Studies* 42, no. 1 (2013): 87–109.

Mugnai, M. *Leibniz's Theory of Relations*. Stuttgart: Meiner, 1992.

Nagel, T. *Mind and Cosmos: Why the Materialist Neo-Darwinian Conception of Nature Is Almost Certainly False*. Oxford: Oxford University Press, 2012.

Parkinson, G. H. R. *Logic and Reality in Leibniz's Metaphysics*. Oxford: Clarendon Press, 1965.

Rescher, N. 'Leibniz on Intermonadic Relations'. In *On Leibniz*, 68–91. Pittsburgh, PA: University of Pittsburgh Press, 2003.

Russell, B. The Principles of Mathematics [1903]. London: George Allen and Unwin, 1937.

Simons, P. 'Metaphysical Systematics: A Lesson from Whitehead'. *Erkenntnis* 48, no. 2/3 (1998): 377–93.

Sprigge, T. L. S. *The Vindication of Absolute Idealism*. Edinburgh: Edinburgh University Press, 1983.

Sprigge, T. L. S. *The God of Metaphysics*. Oxford: Clarendon Press, 2006.

Sprigge, T. L. S. *The Importance of Subjectivity: Selected Essays in Metaphysics and Ethics*, edited by L. B. McHenry. Oxford: Clarendon Press, 2011.

Strawson, G. 'Realistic Monism: Why Physicalism Entails Panpsychism'. In *Consciousness and Its Place in Nature*, edited by Anthony Freeman, 3–31. Exeter: Imprint Academic, 2006.

Strawson, G. *Selves: An Essay in Revisionary Metaphysics*. Oxford: Oxford University Press, 2009.

Ward, J. *The Realm of Ends or Pluralism and Theism: The Gifford Lectures, Delivered in the University of St. Andrews in the Years 1907–1910*. Cambridge: Cambridge University Press, 1911.

Whipple, J. 'The Structure of Leibnizian Simple Substances'. *British Journal for the History of Philosophy* 18, no. 3 (2010): 379–410.

Whitehead, A. N. *Symbolism: Its Meaning and Effect*. New York: MacMillan, 1927.
Whitehead, A. N. The Function of Reason [1929]. Boston, MA: Beacon, 1958.
Whitehead, A. N. Religion in the Making [1926]. New York: Fordham University Press, 1996.

ARTICLE

BRITISH IDEALIST MONADOLOGIES AND THE REALITY OF TIME: HILDA OAKELEY AGAINST MCTAGGART, LEIBNIZ, AND OTHERS[1]

Emily Thomas

In the early twentieth century, a rare strain of British idealism emerged which took Leibniz's *Monadology* as its starting point. This paper discusses a variant of that strain, offered by Hilda Oakeley (1867–1950). I set Oakeley's monadology in its philosophical context and discuss a key point of conflict between Oakeley and her fellow monadologists: the unreality of time. Oakeley argues that time is fundamentally real, a thesis arguably denied by Leibniz and subsequent monadologists, and by all other British idealists. This paper discusses Oakeley's argument for the reality of time, and Oakeley's attack on the most famous account of the unreality of time offered in her day: that of J. M. E. McTaggart. I show that Oakeley's critique of McTaggart can be extended to challenge *all* monadologists, including that of the great monad, Leibniz himself.

1. INTRODUCTION

In the early twentieth century, the British idealist Hilda Oakeley (1867–1950) argued for a 'personalist' kind of Leibnizian monadology, the starting point of which is the individual perceptions of selves. There is very little literature on Oakeley but this paper argues that she plays an important role in the legacy of Leibniz's *Monadology*. Unlike arguably all of her fellow idealists and monadologists, Oakeley holds that time is fundamentally real. In addition to providing an argument for the reality of time, Oakeley critiques the views of her contemporary J. M. E. McTaggart, that most famous advocate for the unreality of time. This paper argues that Oakeley's critique of

[1] I would like to thank Jeremy Dunham, Pauline Phemister, Bill Mander, and Kris McDaniel for valuable comments on earlier drafts of this article.

McTaggart can be extended beyond the intellectual confines of British idealism to challenge *all* personalist monadologists.

The paper runs as follows. Section 2 details the intellectual context in which Oakeley is working. Section 2.1 introduces British idealist monadologies, a branch of idealism that has been woefully neglected by the history of philosophy. As Oakeley is a little-known figure, Section 2.2 provides a brief introduction to her work. Section 3 considers Oakeley's views on time. Section 3.1 sets out Oakeley's positive argument for the reality of time, and shows that several of her views are still held today. Section 3.2 discusses Oakeley's critique of McTaggart's thesis that our temporal perceptions are misperceptions of an atemporal series. Oakeley persuasively argues that McTaggart has not wholly eliminated time from his account of reality, and that his metaphysics cannot explain our perception of temporal passage. Section 4 shows that Oakeley's critique of McTaggart can be extended to challenge all personalist monadologists, by applying her critique to Leibniz.

2. BRITISH IDEALIST MONADOLOGIES AND OAKELEY

2.1. Flavours of British Idealism

Idealism dominated British philosophy in the late nineteenth century. In the early twentieth century, various anti-idealist 'new realisms' appeared, and during this period disputes between various kinds of idealism proliferated. One way of distinguishing British idealist monadologies from other kinds of idealism is to consider their conflicting accounts of selves, and this distinction is particularly useful to us as it will help us to contextualize Oakeley's views.

British idealism is best known for 'Absolute idealism', the monistic view that the ultimate reality is *one* Absolute consciousness, held for example by F. H. Bradley. However, idealism came in other flavours too. Andrew Seth Pringle-Pattison led the charge for 'personal idealism', arguing that Absolute idealism does not leave room for the individuality of selves. Pringle-Pattison (*Hegelianism and Personality*, 216) argued that reality is pluralist because it comprises *many* minds; although, ultimately, this plurality forms a unified Absolute, such that selves are 'parts of the system of things'. Pringle-Pattison's charge was solidified by a 1902 collective volume *Personal Idealism*, edited by H. G. Sturt. There is a modest body of scholarship on Absolute and personal idealism.[2]

The first decades of the twentieth century saw another flavour of British idealism emerge: 'monadologies'. These idealisms exhibit several elements found in Leibniz's *Monadology*, such as Leibniz's view

---

[2]This includes Cunningham (*Idealistic Argument*), Passmore (*Hundred Years of Philosophy*), Mander (*British Idealism*), and Dunham, Grant, and Watson (*Idealism*).

(AG 214–23)[3] that all monads continually perceive the universe from a unique perspective; that all monads are subject to 'appetition', a kind of nisus that moves the monad from one perception to the next; and that the highest monads take pleasure in pure love. James Ward's 1911 *The Realm of Ends or Pluralism and Theism* provides an early example of British idealist monadology. The monadology of H. Wildon Carr followed a few years later, seemingly independently. Oakeley's monadology emerged later still. There is a modest body of scholarship on Ward, there is a little on Oakeley, and there is none on Carr (with the exception of Oakeley's articles).[4]

As there is reason to believe that Oakeley draws on Carr's monadology, I give a brief overview of it. Carr's 'Philosophy as Monadology' argues that philosophy is the study of living experience, and centres of living experience are monads. For Carr ('Philosophy as Monadology', 125–30) a monad is anything that is the subject of experience. Carr places heavy emphasis on the unique perspectival experience of monads, as illustrated here by his metaphor of passengers sitting on a train:

> Each of my fellow-passengers is, like myself, a mind. Each mind is a universe, a universe reflected into a centre as though into a mirror, and every centre is an individual point of view. Between one mind and another there is absolutely nothing in common ... To a mind all reality is experience, and to each mind its own experience.
>
> (Carr, 'Philosophy as Monadology', 127)

For Carr ('Philosophy as Monadology', 133) monads lack windows in the sense that they cannot 'directly' communicate with one another – by which I read Carr as meaning that no monad can inhabit the unique perspective of another – but they can communicate 'indirectly' through body language and speech.

McTaggart's two-volume magnum opus *Nature of Existence* argues that the universe is comprised exclusively of selves and their perceptions. McTaggart has been understood as a personal idealist and as a monadologist. McTaggart (*Nature of Existence I*, 50) states that his idealism is that of Berkeley, Leibniz, and Hegel, in that (as McTaggart reads them) nothing exists except for spirit. McTaggart's system displays a number of Leibnizian hallmarks.[5]

---

[3]'AG' references *G. W. Leibniz: Philosophical Essays*.

[4]On Ward, see Cunningham (*Idealistic Argument*, 169–201), Murray (*Philosophy of James Ward*), Basile (*Metaphysics of Causation*, 32–62) and Dunham, Grant, and Watson (*Idealism*, 175–89). On Oakeley, see Thomas ('Oakeley on Idealism'). Our monadologists also make very brief appearances elsewhere: Passmore (*Hundred Years of Philosophy*, 301) mentions Carr; and Mander (*British Idealism*, 368, 533) mentions Carr and Oakeley.

[5]Some examples. McTaggart holds that every part of the universe reflects every other part (*Nature of Existence I*, 299); he emphasizes the importance of love (*Nature of Existence II*, 147); and – controversially – he reads Leibniz as holding that substances are infinitely divisible and advocates the same view (*Nature of Existence I*, 189).

These have been noted in the literature[6] and their origins puzzled over. Mander (*British Idealism*, 372) suggests that the Leibnizian elements in McTaggart may be evidence of the influence of Ward, who tutored McTaggart at Cambridge. However, Ward's correspondence suggests that he was not friendly with McTaggart,[7] casting some doubt on this line of influence. McTaggart's work may have taken on Leibnizian elements independently, as he lectured on Leibniz. Another possible line of influence is McTaggart's colleague (and, at one time, friend) the 'new realist' Bertrand Russell, who lectured on Leibniz at Cambridge in 1899 in place of McTaggart (while McTaggart was away), and subsequently published a study of Leibniz.

British idealist monadologists disagree with absolute idealists *and* with personalist idealists over the nature of selves. To illustrate, Pringle-Pattison (*Idea of God*, 257–9) criticizes the 'old doctrine of the soul-substance as a kind of metaphysical atom', and – picking out McTaggart for special attention – argues that it is absurd to talk of selves existing in their own right, as selves exist as an 'organ' of the Absolute. Presumably, Pringle-Pattison would subject the systems of Ward, Carr, and Oakeley to the same criticisms. Against the Absolute idealists, monadologists agree with personal idealists of Pringle-Pattison's ilk that selves are real to a significant degree. They are disagreeing over *how* real selves are: personal idealists conceive selves as somewhat real but ultimately unify them in the Absolute, whereas monadologists conceive selves as absolutely real, with no Absolute in sight. The Leibniz, one might say, is in the details.

Another way of distinguishing monadologies from other kinds of idealism lies in the texts they are drawing from. Tim Crane ('Philosophy, Logic, Science, History', 23) has argued that we should understand a philosophical tradition as a collection of interrelated texts, rather than as a body of doctrines or techniques. This thesis can be applied here: Leibnizian idealisms can be distinguished from their fellows in virtue of the fact they are not drawing on (or at least, not *only* drawing on) Hegel's *Encyclopaedia* or Spinoza's *Ethics*; instead, they are drawing on Leibniz's *Monadology*. This textual context *grounds* the Leibnizian hallmarks found in British idealist monadologies, such as the independent reality of selves, or the value of love.

British idealism is frequently known as 'British Hegelianism' for a good reason: many prominent British idealists – including William Wallace, T. H. Green, Bradley, Bernard Bosanquet, Pringle-Pattison, and McTaggart – draw frequently and explicitly on Hegel. However, there are reasons to be

---

[6]See Broad (*Examination of McTaggart's Philosophy I*, 7) and Geach (*Truth, Love and Immortality*, 17). Basile ('Idealist Hydra', 996) goes farther than others in arguing that McTaggart's 'Leibnizian' philosophy defends a system of 'interconnected monads'.

[7]In a 1907 letter to G. H. Howison, Ward writes, 'I do not see much of McTaggart, he regards me as a hopeless old fossil: I regard him as a wild *a priori* dreamer, too quaint and too heedless of facts' (BANC MSS C-B 1037, George Holmes Howison Papers, Bancroft Library, UC Berkeley). Thanks to Jeremy Dunham for pointing me to this.

unhappy with this label. In addition to downplaying the influence of thinkers such as Spinoza and Kant, the monistic connotations of the 'British Hegelianism' label excludes the radically pluralist metaphysics of British idealist monadologists, who are drawing partially or exclusively on Leibniz.

2.2. Introducing Hilda Oakeley

As Oakeley is a little-known figure, it will be helpful to give some biographical information. Oakeley came up to Oxford in 1894, studying for her undergraduate degree under the Absolute idealists Wallace, Bosanquet, and Edward Caird. (Whilst Oakeley studied for a degree she was not awarded one, as Oxford did not award degrees to women until 1920.) After teaching philosophy at McGill University and the University of Manchester, Oakeley joined the philosophy department at King's College London in 1907. She stayed at King's for the rest of her career and, by 1940, her status was such that she was elected President of the Aristotelian Society.[8] Oakeley was prolific and her writings range from the philosophy of history to politics. We will be focusing on her metaphysics.

Oakeley begins characterizing her system as a monadology from the 1920s.[9] Oakeley's work particularly draws on Leibniz's conception of selves but it also bears other Leibnizian hallmarks. For example, Oakeley ('Symposium', 309) holds that the highest possible value is the love of selves for other selves. In a way that is reminiscent of Leibnizian appetition, Oakeley (*Philosophy of Personality*, 16) further argues that selves are continually striving; for Oakeley, this is an effort to overcome the limits of our perspectives. I suggest that the Leibnizian elements in Oakeley's work are at least partly evidence of her collegial relationship with Carr. Oakeley writes of Carr with personal warmth[10] and she was extremely familiar with his philosophy: they worked together at King's College London from 1914 to 1925; she wrote on his monadology, see Oakeley ('Discussion of Dr Wildon Carr'); and she wrote a memorial piece on Carr and his work after his death, see Oakeley ('In Memoriam'). Additionally, in the context of setting out her own monadology, Oakeley (*Philosophy of Personality*, 29–30) writes approvingly that, amongst modern Leibnizian philosophies, Carr 'seems to come nearest' to Leibniz.

---

[8] For more on her career, see Oakeley's (*My Adventures in Education*) autobiography and Howarth ('Oakeley, Hilda Diana (1867–1950)').

[9] On how Oakeley's monadology fits into her wider idealist views, see Thomas ('Oakeley on Idealism').

[10] For example, Oakeley ('In Memoriam', 258) describes Carr's 'undimmed intellectual energy' and his 'urbanity in debate'.

3. OAKELEY AND THE REALITY OF TIME

3.1. Oakeley's Argument for the Reality of Time

Oakeley's *Philosophy of Personality* argues that reality is fundamentally temporal. Her argument is not given in premise form but, on my reading, I reconstruct it as follows:

(i) Selves are 'personalist' monads
(ii) Selves perceive temporal passage
(iii) The best explanation for selves' perception of temporal passage is that reality is fundamentally temporal
(iv) Reality is fundamentally temporal

The argument is valid and it would remain so even if (i) were removed, as (ii) and (iii) together are sufficient to establish (iv). Although, logically speaking, (i) is superfluous, I have included it because it will help us to understand the flow of Oakeley's reasoning.

I will discuss each premise in turn, beginning with (i). As the title of *Philosophy of Personality* suggests, Oakeley is deeply concerned with 'persons'. In the preface, Oakeley (*Philosophy of Personality*, 7) describes personality 'as that principle which must needs endow experience with individuality'. In this context, she mentions McTaggart's ('Personality', 773) view that a 'self' is identical with 'personality', and there is no indication that Oakeley departs from this. A little later, Oakeley (*Philosophy of Personality*, 21–2) explains that 'personality appears wherever there is consciousness or awareness of reality'. For Oakeley, a personality – a person – is an individual, conscious self.[11]

Oakeley (*Philosophy of Personality*, 26–7) states that a philosophy of persons such as hers must go back to Leibniz's *Monadology* as one of its original sources, and that the *Monadology* has been interpreted in two ways. The first interpretation emphasizes the relationship between monads and the divine monad from which they proceed: this tends towards monism, encouraging us to conceive monads as 'fulgurations' of God. This is a reference to a passage in Leibniz's *Monadology* which states that all created monads are generated by 'continual fulgurations of the divinity' (AG 219).[12] In contrast, the second interpretation emphasizes the monads' 'exclusive individuality in respect to the perspective theory of knowledge', the unique epistemic perspective of each monad: this tends towards

---

[11] For Oakeley, all monads are selves. In contrast, Leibniz (AG 214-5) allows for living monads that do not possess the full consciousness of minds.

[12] A 'fulguration' is literally a lightning flash. This statement could be read to mean (as Oakeley implies) that monads are parts or emanations of the divine; or merely that monads are sustained by the divine. Blank ('Substance Monism and Substance Pluralism') provides a recent discussion on whether Leibniz is best read as a monist or pluralist.

pluralism, focusing on the plurality of 'worlds' which are perceived from the individual point of view of the knower. On the first 'monistic' monadology, selves are fulgurations, dependent on something larger; on the second 'personalist' monadology, selves are independent individuals.

Oakeley argues that Leibniz's system is not wholly personalist, and as such he did not develop his system to the fullest. However, the thinkers of her period are now in a position to do so:

> Perhaps it required the movement of thought from Kant, through Hegel and later Idealism ... to bring about the definite and acute insight of modern idealists into the truth, that the individual is integrally bound up with his world, that if his outlook upon reality is unique and incommunicable, so is his world only for himself in its total nature.
>
> It is then in the philosophy of Leibniz ... that may be found the starting-point of the development of the personalistic interpretation.
>
> (Oakeley, *Philosophy of Personality*, 28–9)

Oakeley places Leibniz in the idealist tradition[13] and the starting point of her metaphysics is the Leibnizian insight that individuals are 'bound up' with their unique worlds. This can be illuminated by a discussion in McTaggart, who considers the view that perception is the awareness of 'sense-data', mental objects that minds perceive directly, such as noises or colours. McTaggart ('Personality', 774–5) explains that it is commonly held that 'that which falls wholly within a mind' – such as sense-data – is *not* perceptible by any mind except that in which it falls, entailing that each sense-datum can only be perceived by one person.[14] Although Oakeley does not put the point in the language of sense-data, the spirit of her view is the same: a self's perspective, or 'world of knowledge', is unique to it. Oakeley favours a personalist monadology over a monistic one for many reasons, not least because (in agreement Carr and McTaggart) she thinks it best accords with our experience as conscious selves, and with the value we wish to place on selves.

Although (i) is not logically necessary to establishing (iv), the argument has special force if (i) is included. This is because, if selves are conceived as personalist monads, then the starting point of one's metaphysics is the perception or experience of selves. Any philosophic position that starts from selves' perception must *account* for selves' perception. As Oakeley ('On the Meaning of Value', 435) puts it, 'the nature of the real must be such as to account for the facts of experience'. This does not commit the

---

[13]This is controversial. For reasons to reject idealist readings of Leibniz, see Loptson ('Was Leibniz an Idealist?'), Phemister (*Leibniz and the Natural World*), Garber (*Leibniz: Body, Substance, Monad*), and Arthur ('Leibniz's Theory of Time').

[14]Although McTaggart considers sense-data, he ultimately denies their reality; see McTaggart (*Nature of Existence II*, 57).

monadologist to asserting the truth of everything we perceive, but it does commit them to satisfactorily explaining everything we perceive. If the monadologist accepts that we perceive temporal passage but holds that this is a misperception, they must explain that misperception.

Let us move on to (ii). Oakeley (*Philosophy of Personality*, 47) claims that selves have a particular kind of experience or perception:

> the quality of experience as arising from the unknown, emerging as the novel, and thence from firm settlement as actuality, passing to the irrevocable past, or the coming to birth of the object, its growth to ripeness, and old age.

In today's parlance, Oakeley is describing our experience of 'temporal passage', the movement of time: our sense of an event arising from the 'unknown' future, becoming present 'actuality', and passing into the 'irrevocable' past. Many philosophers have claimed we experience temporal passage.[15] Even D. C. Williams's ('Myth of Passage', 466) impassioned critique of passage claims that it is 'futile' to deny our experience of it, for we are 'immediately and poignantly involved in the jerk and whoosh of process, the felt flow of one moment into the next'.

Oakeley goes on to argue for (iii):

> [The doctrine that reality] is non-temporal, must be sacrificed if we are to gain an intelligible view of the character of our actual knowledge. The interpretations of absolute Idealism, of Spinozism, even that of Leibniz, though he allows the time sequence to be a *phenomenon bene fundatum* are proved to be in some way imperfect as instruments for making this intelligible. For there results from this type of interpretation a conception which ... [is] incapable of justifying certain characters of our knowledge experience.
> (Oakeley, *Philosophy of Personality*, 46–7)

Oakeley is arguing that those who hold reality to be non-temporal – such as the absolute idealists; Leibniz; and Spinoza – cannot account for our actual knowledge, our perception of temporal passage. Oakeley seems to have held this view from her earliest work.[16]

In defending the doctrine that reality is temporal, I take Oakeley to mean that temporal passage is fundamentally real. (My addition of the qualifier 'fundamental' merely implies that time passes on the rock-bottom level of reality, as opposed to say having some degree of reality as a well-founded appearance.) In holding that temporal passage is real, and events really change their temporal properties from future to present to past, Oakeley is an 'A theorist'. This terminology has its roots in McTaggart, who

---

[15]Recent examples include Craig (*Tensed Theory of Time*, 138) and van Inwagen (*Metaphysics*, 64).

[16]Oakeley's ('Reality and Value', 99) states that, given the capacity of the mind to apprehend things as they really are, time cannot be an illusion. Frustratingly, the paper does not elaborate.

distinguishes two ways of ordering events. On the 'A series', events are ordered in virtue of possessing temporal properties, such as being 'past'. In contrast, on the 'B series', events are ordered by being 'earlier' or 'later' than other events. Arguments from our experience of temporal passage to A theory can still be found in metaphysics today, and – whilst controversial – they are generally acknowledged to have force.[17]

Oakeley concludes (iv) that reality is temporal. In the British idealist context, this position is nothing less than radical. As far as I am aware, *every* single British idealist except Oakeley holds that in some sense time is unreal.[18]

3.2. Oakeley's Critique of McTaggart

Carr and McTaggart accept the theses expressed by (i) and (ii) but not (iii). In her later work, Oakeley critiques McTaggart's rejection of (iii), and that is the subject of this section.

Oakeley really begins to engage with McTaggart's work from the late 1920s, and she expresses high praise it. For example, Oakeley ('Symposium', 309) praises McTaggart's Leibnizian view that the supreme value of experience is love of selves for other selves; Oakeley (*History and the Self*, 256–7) applauds McTaggart's advocacy of the importance of persons; and Oakeley (*History and the Self*, 20–21) writes that McTaggart's 'genius' philosophy of selves has been of great value to her work, even though she finds his timeless universe 'untenable'. Just how untenable will become apparent shortly.

McTaggart is best known for his 1908 argument for the unreality of time. As this argument is not our focus, I will recount it very briefly.[19] McTaggart ('Unreality of Time', 467–70) argues that there is no time without the A series, because only the A series provides change in temporal properties. However, McTaggart argues that the A series is contradictory – if an event is present it cannot also be past and future – and so the A series cannot be true of reality. As there is no time without the A series, time is an unreal appearance.

---

[17]For recent discussion, see Callender ('Common Now') and Skow ('Experience and the Passage of Time').

[18]For example, Bradley (*Appearance and Reality*, 205–22) argues that time is mere appearance. In a symposium, Bosanquet and – perhaps surprisingly – the soon-to-be 'new realist' G. E. Moore agree that time is ultimately unreal (Bosanquet, Hodgson, and Moore, *In What Sense, If Any, Do Past*). Pringle-Pattison (*Idea of God*, 343–66) argues that time is entirely dependent on the succession of our experienced content, such that without minds, there would be no time; ultimate reality transcends time.

[19]The ample literature on this argument includes Mander ('McTaggart on Error and Time') and Dainton (*Time and Space*, 13–26). On McTaggart's work more generally, see Broad (*Examination of McTaggart's Philosophy I*; *Examination of McTaggart's Philosophy II*), Geach (*Truth, Love and Immortality*), Mander (*British Idealism*, 369–76) and McDaniel ('John M. E. McTaggart').

Taking himself to have established the unreality of time, McTaggart goes on to argue – most fully in *Nature of Existence* – that our temporal 'perceptions' are misperceptions of an atemporal reality. Like Oakeley, McTaggart agrees that the nature of the real must account for the facts of experience. McTaggart (*Nature of Existence I*, 50–1) explains that his methodology is specifically concerned with relating the characteristics of our experience to the general nature of the existent. With this in mind, McTaggart (*Nature of Existence II*, 194) sets out to reconcile appearances with the real nature of the universe, and one of these appearances is time.

McTaggart holds that our misperceptions of things as being in the A series – as being present, future, or past – provide clues to the underlying nature of reality:

> [T]he misperception which gives us the A series clearly implies that the terms which are misperceived as forming it, do really form a series ... though not a time-series. The fact that the terms are in such a series involves that each term has a definite position on one side or the other of any given term, and is either nearer to it or further from it than any other term on the same side of it.
> (*Nature of Existence II*, 213)

McTaggart labels this real series the 'C series'.

McTaggart (*Nature of Existence II*, 213) argues that the relation which connects terms in the C series must share various characteristics of the relation 'earlier than', such as being transitive and asymmetric.[20] McTaggart concludes that this relation is the 'inclusion' relation:

> Of any two terms in the B series, one is earlier than the other, which is later than the first, and by means of these relations all the terms can be arranged in one definite order. And of any two terms in the C series, one is included in the other, which includes the first, and by means of these relations all the terms can be arranged in one definite order ... it is the relations of "included in" and "inclusive of" which appear as the relations of "earlier than" and "later than".
> (*Nature of Existence II*, 240)

There is a precise parallelism between this 'inclusion series' and the things we misperceive as the time series, and for this reason – echoing Leibniz's remarks on extension and motion (see below) – McTaggart (*Nature of Existence II*, 273) describes time as a 'phenomenon bene fundatum'. The idea is that we are perceiving 'through a glass, darkly': the murky nature of the lens leads us to perceive reality as temporal when it is not, but our misperceptions are founded in reality. McTaggart (*Nature of Existence II*, 365) further

---

[20]Relation R is 'transitive' iff, if a is related by R to b, and b is related by R to c, then a is related by R to c. Relation R is 'asymmetric' iff, if a is related by R to b, then b is not related by R to a.

argues that the C series possesses one stage that the time series does not: a 'final stage' that will include all the contents of the preceeding stages. When selves reach the final stage, they will perceive the whole series which *sub specie temporis* is their life throughout time, although they will 'correctly' perceive these stages as not really being in time (McTaggart, *Nature of Existence II*, 389). McTaggart's work provides one way of rejecting (iii): by arguing that our temporal perceptions are misperceptions of the C series, McTaggart provides an alternative explanation for our experience of time.

Oakeley is impressed by McTaggart's attempted parallelism between appearance and reality[21] but she firmly rejects it. Oakeley argues that McTaggart's account of reality does not account for our experiences, and thus McTaggart's system fails on its own terms. In a series of articles, Oakeley presents a battery of arguments against McTaggart's views on time.[22] I will discuss an argument presented in Oakeley's 1930 paper 'Time and the Self in McTaggart's System'. This argument poses an objection to McTaggart's inclusion relation, and I read it as having two prongs: our notion of 'inclusion' depends on the prior notion of time, and 'inclusion' does not account for our perception of temporal passage. I will discuss them in turn.

The first prong runs as follows. An anti-realist about time must provide a satisfactory explanation of reality that is wholly free of time: it would be contradictory to deny the fundamental reality of time *and* employ temporal notions at the fundamental level of ontology. Oakeley argues that McTaggart is guilty of this contradiction, because 'inclusion' is an inherently temporal notion:

> There would be no meaning in a statement that A is included in B unless behind the statement there lay the experience, actual or possible, of the occurrence of one after the other. How otherwise is the difference between the first

---

[21] Oakeley ('Philosophy of Time and the Timeless', 105) describes McTaggart's parallelism as 'unique' and 'daring'. Broad (*Examination of McTaggart's Philosophy II*, 787) agrees that philosophers who deny time must account for its appearance, and adds: 'But how completely most of them have shirked this job, and how well has McTaggart done it!' Interestingly, Mander ('McTaggart on Error and Time', 162) argues that McTaggart's parallelism poses a 'direct challenge' to the dominant British idealist position that time is an appearance of the Absolute. Absolute idealism offers no parallelism; it is hard to see how a monistic Absolute could admit a series of any kind.

[22] Some examples. Oakeley ('Symposium', 314–5) argues that McTaggart's thesis that the final stage will not contain evil can only be interpreted in two ways – either, evil is included in the final stage and 'is not the evil that it appeared'; or, that which is evil now is 'somehow' the appearance of a real good – both of which are untenable. Oakeley (*History and the Self*, 195–6) argues that an atemporal system of ethics is problematic: 'the irreversibility of the past is a main source of moral experience, involving that in certain respects what has been done *is* irremediable … [hence] the ethical emotion of remorse … need not trouble the unhistoric ethic'. Oakeley ('Philosophy of Time and the Timeless', 127) argues that even *if* time is an illusion generated by the timeless self, McTaggart has not explained how selves would shake that illusion in the final stage.

and the second to be given its value, how otherwise is identity between the two to be avoided? The notion of inclusion is the one selected by McTaggart as the type of the real series, of which the temporal is phenomenal ... But how are we to conceive the relation of the included to the including unless we can think of the terms or events as *now* separated from each other and *then* coming together? Without the temporal form they become identical in inclusion and the series is no longer a series.

(Oakeley 'Time and the Self in McTaggart's System', 183)

To understand Oakeley's argument, it will be helpful to elaborate on it.

As McTaggart would accept, the 'included in' relation holds between non-identical terms. Identity is a symmetric relation, and 'included in' is an asymmetric relation. Given this, if A and B are identical, then one cannot be included in the other. In arguing that the statement 'A is included in B' is meaningless unless the occurrence of one after the other were possible, I read Oakeley as arguing that there are no necessary connections between distinct (i.e. non-identical) existences: if A is really non-identical to B, then A can exist independently or separately of B. For Oakeley, the statement 'A is included in B' entails at time t_1 conceiving A, and *then* at t_2 conceiving A included in B. It is this move – from considering A, *to* A included in B – that leads Oakeley to object that inclusion involves a 'temporal form'.[23]

Oakeley's objection pushes the burden of proof back onto McTaggart, challenging him to explain how this series is not temporal. Several years later, the 'new realist' C. D. Broad – McTaggart's friend and former pupil – published a lengthy study of McTaggart's system. Although there is no indication that Broad read Oakeley, Broad makes an objection that appears to run on similar lines. Broad (*Examination of McTaggart's Philosophy II*, 522–3) asks us to consider a series of propositions on which the preceding propositions logically and asymmetrically entail the next, such that p_1 entails p_2, p_1 and p_2 together entail p_3, and so on. For McTaggart, the terms and relations of this series would be timeless. Against McTaggart, Broad argues that a reference to time 'is essential' because one must know p_1 and p_2 *before* one can know p_3, and one can know p_3 only *after* knowing p_2 and p_1. These arguments from Oakeley and Broad share a common thrust: in conceiving a series, we consider things *before* and *after* one another, and thus a series involves time.

On the second prong, Oakeley objects that the inclusion relation (whether or not it is temporal) cannot account for our perception of temporal passage:

The difficulty which McTaggart seems to experience in determining what type of real series it is of which the time-series is a misperception, may suggest doubts

---

[23] Against McTaggart's thesis that selves *progress* through the stages of their lives, Oakeley ('Philosophy of Time and the Timeless', 119–20) makes the related objection that progress is an 'essentially temporal' notion, and involving such notions in the C series 'threatens a grave distortion' of McTaggart's philosophic framework.

> whether the notion of a series is the true logical equivalent of the process of time. To some thinkers, the nature of time has appeared to be better indicated by the idea of a continuous passage which is discrete in its moments only in relation to our mode of conceiving it, stages in the passage being distinguished primarily on account of their importance whether in view of practice or in view of the systematic and intelligible conception of the world.
> (Oakeley, 'Time and the Self in McTaggart's System', 186)

Oakeley is arguing here that the notion of a series is not the logical equivalent of the process of time, and a better characterization of time is 'continuous passage'. I will discuss each element of this characterization in turn.

For McTaggart, time is well founded in the C series, which comprises discrete perceptions. Oakeley is arguing that time is better characterized as *continuous*; as I read her, this is because our perception of temporal passage is continuous. Elsewhere, Oakeley ('Status of the Past', 243–4) places great emphasis on the 'continuity of self-consciousness'. As our perception of passage is continuous, a discrete C series cannot account for it. Additionally, McTaggart's C series does not contain *passage*, and so it is hard to see how it can explain our perception of passage. Broad (*Examination of McTaggart's Philosophy II*, 546) also make this objection: 'But where, in all this timeless co-existence of non-temporal series, can the appearance of the *passage* of A-characteristics from one term to another arise?'[24] For Oakeley, McTaggart's C series is not parallel to our experience of time because it lacks continuity and passage.

The two prongs of Oakeley's critique come together to show that McTaggart's account of temporal 'misperception' is unsatisfactory: McTaggart has not eliminated time from the C series, and the C series cannot account for our perception of continuous passage. This entails there is some flaw in McTaggart's logical argument for the unreality of time. As Oakeley ('Philosophy of Time and the Timeless', 116) puts it, 'That the contradiction which McTaggart finds in the nature of time is less than that which results from the postulate of its unreality seems suggested by his own work.' As McTaggart passed away in 1925, prior to Oakeley's critiques, we do not know what he might have made of them. At the conclusion of the last of her critiques, Oakeley writes:

> Finally, I would venture to express a surmise – mainly the result of reflection on McTaggart's philosophy of time – that an a priori proof of the reality of time is not inconceivable. The line it would follow would proceed from the assumption that the 'error' must be attributed to the subject ... we may accept the view that time is in the self *without* holding consciousness of time to be error. Must there not be postulated sufficient harmony between

---

[24]Broad adds a sentiment that Oakeley would certainly accept: 'I cannot help thinking that there could be no *appearance* of becoming *anywhere* unless there were *real* becoming *somewhere*.'

the self and its experienced object, to assure that so universal and inescapable a form must be in things, and events, as well as in the mind?
(Oakeley 'Philosophy of Time and the Timeless', 126–7)

McTaggart and Oakeley agree that selves can be understood as Leibnizian monads but, against McTaggart, Oakeley holds that temporal passage – that 'universal and inescapable' form – is not an error generated by selves but a feature of reality.

4. EXTENDING OAKELEY'S CRITIQUE BEYOND MCTAGGART

This section argues that Oakeley's critique of McTaggart can be extended to challenge *all* monadologists who deny the reality of time. The challenge is this: If time is merely a well-founded phenomenon, then the grounds of the phenomenon must be purely non-temporal and such as to account for our perception of temporal passage. The latter part of this challenge renders it especially applicable to personalist monadologists because, for these theorists, the starting point of metaphysics is the perception or experience of individual selves. Given the importance monadologists place on perception, they should satisfactorily account for a perception as all-pervasive as time. To demonstrate the force of Oakeley's challenge, I show how it can be applied to the work of the great monadologist, Leibniz himself.

A full discussion of Leibniz's account of time would reach into his wider metaphysics and intellectual context, and there is not space to attempt that here. Instead, I confine myself to sketching Leibniz's pertinent views. Leibniz's *Primary Texts* (c. 1686) states, 'Extension and motion, as well as bodies themselves ... are not substances, but true phenomena, like rainbows' (AG 34). A rainbow is a phenomenon that is 'well founded' in real beings, its water droplets; similarly, material bodies, spatial extension and movement through time are well-founded phenomena grounded in monads. Oakeley ('Time and the Self in McTaggart's System', 175) briefly states that she perceives 'important affinities' between the views of Leibniz and McTaggart on time, and one of these affinities is likely their use of well-founded phenomena.

In his later work, Leibniz describes time as 'ideal'. For example, in a 1705 letter, Leibniz writes:

[Time] is nothing but a principle of relations, a foundation of the order of things, in so far as one conceives their successive existence, or without which they would exist together. It must be the same in the case of space ... Both of these foundations are true, although they are ideal.
(trans. Hartz and Cover, ' Space and Time in the Leibnizian Metaphysic', 501)

In a 1716 letter, Leibniz writes that the certain order that is time is analogous to genealogical relations, which express real truths yet are 'ideal things' (AG

339). Leibniz is arguing that genealogical relations hold between brothers, sisters, and mothers and yet they are merely ideal things, *entia rationis*, or objects of the mind; similarly, temporal relations hold between bodies and yet are merely ideal.

Scholars are deeply divided over how to understand Leibniz's account of time.[25] One dispute is over whether time is directly founded in monads. For example, drawing on Leibniz's thesis that monads continually perceive, Rescher (*Leibniz – An Introduction to his Philosophy*, 90–1) argues that time is grounded in the succession of monads' changes of state, from one perception to another. On Rescher's reading, Leibniz holds a two-tier ontology: monads lie on the bottom tier of reality, and time lies on the tier above. In contrast, Hartz and Cover ('Space and Time in the Leibnizian Metaphysic', 512) have argued against Rescher that, in his mature writings, Leibniz conceives matter as a well-founded phenomenon and time is a mere 'ideal order', abstracted away from what is well founded. On this reading, Leibniz's mature writings posit a three-tier ontology: monads lie at the foundational level; well-founded phenomena grounded in monads, such as material bodies, lie one tier up; and ideal things, such as time, lie two tiers up.

Another scholarly dispute is over whether monads are *in* time. On one line of interpretation, monads (directly or indirectly) ground time but they are not in time. As we saw above, Oakeley reads Leibniz in this way. More recently, McDonough ('Leibniz's Philosophy of Physics', §5.2) writes of Leibniz's mature metaphysics that whilst Leibnizian bodies – understood here as well-founded phenomena – stand in temporal relations to one another, monads do not. On another line of interpretation, monads ground time and they are themselves related in time. Arthur (*Leibniz*, 159–65) provides a recent defence of this reading, specifically rejecting the view that Leibniz anticipates McTaggart, and arguing that monads really succeed one another in time.

The first prong of Oakeley's challenge holds that, if time is merely a well-founded phenomenon, its ground must be purely non-temporal. If Leibniz conceives monads to be in time, there is no work for this prong to do (as Leibniz would agree with Oakeley that fundamental reality is temporal). However, if Leibniz conceives monads to be outside of time, then there is ample work for this prong of Oakeley's challenge. I will enter deeper into this latter reading of Leibniz, and show how Oakeley's challenge would apply to it.

Let us assume that Leibnizian monads are not in time. On this position, monads would still undergo a succession of states. In response to such a position, the challenge's first prong would object that the fundamental level of reality is not pure from time. Just as Oakeley objected that McTaggart's notion of inclusion depended on the prior notion of time, so it could be objected that Leibniz's notion of succession depends on the prior notion

---

[25]Recent overviews of the scholarship are Lloyd ('Situating Time in the Leibnizian Hierarchy') and McDonough ('Leibniz's Philosophy of Physics').

of time. There is a case to be made that the Leibnizian process whereby a monad changes its state, from having perception A to perception B, is inherently temporal: at time t_1 the monad has perception A, and at t_2 it has perception B.

Confirmation of the force of this worry is that it has already been raised independently in the Leibniz scholarship. For example, Russell (*Critical Exposition of the Philosophy of Leibniz*, 51–3) reads Leibniz as attempting to eliminate time from monads or substances, but argues that this elimination cannot be effected because a substance has a state at one moment, and not at the next. Russell concludes that time is necessarily presupposed in Leibniz's treatment of substance, and the fact it is denied in the conclusion 'is not a triumph, but a contradiction'. More recent scholars agree that succession commits Leibniz to temporality.[26] It is hard to see how Leibniz could deny that succession involves time, as succession seems to involve monads being *now* in one state, and *then* at another. One possible strategy is found in Arthur ('Leibniz's Theory of Time', 276), who argues that we need not consider a monad as having different states at different moments of time, instead the states of a monads can be distinguished by their compatibility with other states. Even if this strategy could be made to work, at the very least Oakeley's challenge puts pressure on the thesis that succession does not depend on time.

The second prong of Oakeley's challenge argues that the grounds of the phenomenon of time must account for our perception of temporal passage. This prong has force against Leibniz even if Leibnizian monads are in time because there are compelling (though not irrefutable) reasons to read Leibniz as holding that the perceptual lives of monads are discrete: they ground matter, which is certainly discrete;[27] and Leibniz speaks of perceptions following one another in a seemingly discrete way.[28] Assuming that Leibniz holds perception to be discrete, Oakeley could argue that because our perception of passage is *continuous*, successive change in discrete monads' states cannot ground time.

Additionally, it could be argued that succession alone cannot account for our perception of temporal *passage*. Whilst critiquing McTaggart, Broad (*Examination of McTaggart's Philosophy II*, 187) writes that temporal passage, or 'becoming', poses a general objection to *all* attempts to 'get rid' of time. Writing on Leibniz, Arthur ('Leibniz's Theory of Time', 276–7) argues that although a mere series of monadic states need not be temporal, by itself this supposition cannot account for the way that monads *pass*

---

[26]For example, McGuire ('"Labyrinthus Continui"', 315) writes, 'Since states exist at one moment and not at the next, activity is irredeemably temporal.' Frankel (*Leibniz on the Foundations of Space and Time*, 93) agrees.

[27]For example, Leibniz writes, 'Matter appears to us to be a continuum, but it only appears so' (trans. Hartz and Cover, 'Space and Time', 501).

[28]On this, see Anapolitanos (*Representation, Continuity and the Spatiotemporal*, 134–52) and Phemister (*Leibniz and the Natural World*, 142–9).

from one state to another. One way of arriving at passage – suggested in Arthur ('Leibniz's Theory of Time', 276–7) – would be to connect Leibnizian appetition with becoming, a principle of temporal change. Of course, this strategy would only be available on the assumption that monads are in time; to hold that monads are subject to appetition conceived as temporal change, and yet deny that monads are in time, would be difficult in the extreme.

Oakeley's advocacy of the reality of time provides her with a unique position amongst British idealists, and her critique of McTaggart offers a challenge to monadologists that extends far beyond her historical milieu. By extending Oakeley's critique of McTaggart to Leibniz, this discussion has demonstrated the force of Oakeley's challenge. Any monadologist who takes the perception of monads as their starting point, and denies the reality of time, must explain how the monadic level of reality is pure from time *and* account for our perception of temporal passage.

BIBLIOGRAPHY

Anapolitanos, D. A. *Leibniz: Representation, Continuity and the Spatiotemporal*. Dordrecht, The Netherlands: Kluwer Academic, 1999.
Arthur, Richard T. W. 'Leibniz's Theory of Time'. In *The Natural Philosophy of Leibniz*, edited by Kathleen Okruhlik and James Robert Brown, 263–313. Dordrecht, The Netherlands: D. Reidel, 1985.
Arthur, Richard T. W. *Leibniz*. Cambridge: Polity, 2014.
Basile, Pierfrancesco. *Leibniz, Whitehead and the Metaphysics of Causation*. Basingstoke, UK: Palgrave MacMillan, 2009.
Basile, Pierfrancesco. 'The Idealist Hydra'. *British Journal for the History of Philosophy* 21 (2013): 989–9.
Blank, Andreas. 'Substance Monism and Substance Pluralism in Leibniz's Metaphysical Papers 1675–1676'. *Studia Leibnitiana* 33 (2001): 216–23.
Bosanquet, Bernard, Shadworth H. Hodgson, and G. E. Moore. 'In What Sense, If Any, Do Past and Future Time Exist?' *Mind* 6 (1897): 228–40.
Bradley, F. H. *Appearance and Reality*. London: George Allen & Unwin, 1893.
Broad, C. D. *An Examination of McTaggart's Philosophy I*. Cambridge: Cambridge University Press, 1933.
Broad, C. D. *An Examination of McTaggart's Philosophy II*. Cambridge: Cambridge University Press, 1938.
Callender, Craig. 'The Common Now'. *Philosophical Issues* 18 (2008): 339–61.
Carr, H. Wildon. 'Philosophy as Monadology'. *Proceedings of the Aristotelian Society* 19 (1918–9): 125–46.

Craig, W. L. *The Tensed Theory of Time: A Critical Examination*. Dordrecht, The Netherlands: Kluwer Academic, 2000.
Crane, Tim. 'Philosophy, Logic, Science, History'. *Metaphilosophy* 43 (2012): 20–37.
Cunningham, G. Watts. *The Idealistic Argument in Recent British and American Philosophy*. New York: Century, 1933.
Dainton, Barry. *Time and Space*. Stocksfield: Acumen, 2001.
Dunham, Jeremy, Iain Grant, and Sean Watson. *Idealism: The History of a Philosophy*. Stocksfield, UK: Acumen, 2011.
Frankel, Lois. 'Leibniz on the Foundations of Space and Time'. *Nature and Systems* 3 (1979): 91–8.
Garber, Daniel. *Leibniz: Body, Substance, Monad*. New York: Oxford University Press, 2009.
Geach, Peter. *Truth, Love and Immortality: An Introduction to McTaggart's Philosophy*. Oakland, CA: University of California Press, 1979.
Hartz, Glenn, and J. A. Cover. 'Space and Time in the Leibnizian Metaphysic'. *Nous* 22 (1988): 493–519.
Howarth, Janet. 'Oakeley, Hilda Diana (1867–1950)'. *Oxford Dictionary of National Biography* (2004). Accessed March 1, 2013. http://www.oxforddnb.com/view/article/48502.
van Inwagen, P. *Metaphysics*. Boulder, CO: Westview Press, 2002.
Leibniz, Gottfried. *In G. W. Leibniz: Philosophical Essays*, edited by Roger Ariew and Daniel Garber. Indianapolis, IN: Hackett Publishing Company, 1989.
Lloyd, Rebecca J. 'Situating Time in the Leibnizian Hierarchy of Beings'. *The Southern Journal of Philosophy* 46 (2008): 245–60.
Loptson, Peter. 'Was Leibniz an Idealist?' *Philosophy* 74 (1999): 361–85.
Mander, W. J. 'McTaggart on Error and Time'. *The Modern Schoolman* 75 (1998): 157–69.
Mander, W. J. *British Idealism: A History*. Oxford: Oxford University Press, 2011.
McDaniel, Kris. 'John M. E. McTaggart'. In *The Stanford Encyclopedia of Philosophy*, edited by Edward N. Zalta. Accessed September 30, 2013. http://plato.stanford.edu/archives/win2013/entries/mctaggart/.
McDonough, Jeffrey K. 'Leibniz's Philosophy of Physics'. In *The Stanford Encyclopedia of Philosophy*, edited by Edward N. Zalta. Accessed December 15, 2014. http://plato.stanford.edu/archives/spr2014/entries/leibniz-physics/.
McGuire, J. E. '"Labyrinthus Continui": Leibniz on Substance, Activity, and Matter'. In *Motion and Time, Space and Matter*, edited by Peter K. Machamer and Robert G. Turnbull, 290–326. Columbus: Ohio State University Press, 1976.
McTaggart, J. M. E. 'The Unreality of Time'. *Mind* 17 (1908): 457–74.
McTaggart, J. M. E. 'Personality'. In *Encyclopaedia of Religion and Ethics*, vol. 9, edited by James Hastings, 773–81. Edinburgh: T & T Clark, 1917.
McTaggart, J. M. E. *The Nature of Existence I*. Cambridge: Cambridge University Press, 1921.

McTaggart, J. M. E. *The Nature of Existence II*, edited by C. D. Broad. Cambridge: Cambridge University Press, 1927.
Murray, Andrew. *The Philosophy of James Ward*. Cambridge: Cambridge University Press, 1937.
Oakeley, H. D. 'Reality and Value'. *Proceedings of the Aristotelian Society* 11 (1910–11): 80–107.
Oakeley, H. D. 'Discussion of Dr Wildon Carr's "A Theory of Monads"'. *Proceedings of the Aristotelian Society* 23 (1922–3): 157–72.
Oakeley, H. D. 'On the Meaning of Value'. *The Philosophical Review* 31 (1922): 431–48.
Oakeley, H. D. *A Study in the Philosophy of Personality*. London: Williams & Norgate, 1928.
Oakeley, H. D. '"Symposium: The Principle of Personality in Experience", with John MacMurray'. *Proceedings of the Aristotelian Society* 29 (1928–9): 301–30.
Oakeley, H. D. 'In Memoriam: Herbert Wildon Carr'. *Proceedings of the Aristotelian Society* 31 (1930–1): 285–98.
Oakeley, H. D. 'Time and the Self in McTaggart's System'. *Mind* 39 (1930): 175–93.
Oakeley, H.D. 'The Status of the Past'. *Proceedings of the Aristotelian Society* 32 (1931-2): 227–50.
Oakeley, H. D. *History and the Self*. London: Williams & Norgate, 1934.
Oakeley, H. D. *My Adventures in Education*. London: Williams & Norgate, 1939.
Oakeley, H. D. 'The Philosophy of Time and the Timeless in McTaggart's Nature of Existence'. *Proceedings of the Aristotelian Society* 47 (1946–7): 105–28.
Passmore, John. *A Hundred Years of Philosophy*. London: Gerald Duckwork, 1957.
Phemister, P. *Leibniz and the Natural World: Activity, Passivity, and Corporeal Substances in Leibniz's Philosophy*. Dordrecht, The Netherlands: Springer, 2005.
Pringle-Pattison, Andrew Seth. *Hegelianism and Personality*. Edinburgh, UK: William Blackwood, 1887.
Pringle-Pattison, Andrew Seth. *The Idea of God*. New York: Oxford University Press, 1917.
Rescher, Nicholas. *Leibniz – An Introduction to His Philosophy*. Totowa, NJ: Rowman and Littlefield, 1979.
Russell, Bertrand. *A Critical Exposition of the Philosophy of Leibniz*. 2nd ed. London: Allen and Unwin, 1937 [1900].
Skow, Bradford. 'Experience and the Passage of Time'. *Philosophical Perspectives* 25 (2011): 359–87.
Thomas, Emily. 'Hilda Oakeley on Idealism, History and the Real Past'. *British Journal for the History of Philosophy* 23 (Forthcoming): 5.
Williams, D. C. 'The Myth of Passage'. *Journal of Philosophy* 48 (1951): 457–72.

Article

Heidegger on the Being of Monads: Lessons in Leibniz and in the Practice of Reading the History of Philosophy[1]

Paul Lodge

> This paper is a discussion of the treatment of Leibniz's conception of substance in Heidegger's *The Metaphysical Foundations of Logic*. I explain Heidegger's account, consider its relation to recent interpretations of Leibniz in the Anglophone secondary literature, and reflect on the ways in which Heidegger's methodology may illuminate what it is to read Leibniz and other figures in the history of philosophy.

In 1928 Heidegger gave the lecture course *The Metaphysical Foundations of Logic* (MFL). Part of the course, which was published by Heidegger in 1967 in his collection *Pathmarks* (P), contains an interpretation of Leibniz's conception of substance.[2] In this paper, I will discuss this component and consider its relationship to some prominent accounts of Leibniz on substance that have appeared in recent Anglophone scholarship.[3] I will also consider whether there are methodological lessons that Anglophone scholars might learn from Heidegger by considering the hermeneutic principles that he employs.

---

[1]Thanks to the following people for helpful comments on earlier drafts of the paper: Mike Beaney, Jeremy Dunham, Mogens Laerke, David Leopold, Stephen Mulhall, Pauline Phemister, Lloyd Strickland, John Whipple, and an anonymous referee for the journal.

[2]The version in *Pathmarks* is entitled 'From the Last Marburg Course'. The differences are mainly due to Heidegger's omissions from MFL in P. However, there are some additions and I shall draw on both versions below.

[3]It should be noted I will be concerned here only with the ways in which a reading of Heidegger's lectures might be brought into dialogue with the reception of Leibniz in recent Anglophone literature. Furthermore, the discussion is aimed primarily at those who are unfamiliar with Heidegger's lectures. There would be much more to be said were considerations of the reception of Heidegger's account of Leibniz in non-Anglophone secondary literature included.

1. METHODOLOGICAL ISSUES

At the beginning of *Pathmarks*, Heidegger offers an account of why he chose to lecture on Leibniz in 1928, telling his readers that the 'interpretations were shaped by the insight that in our philosophical thought we are in dialogue with a thinker of previous times', where 'this means something other than completing a historiographical presentation of philosophy's history' (GA9 373/P63).

We can see from this that Heidegger did not take himself to be engaging in what Robert Sleigh Jr. has called 'exegetical history' (*Leibniz and Arnauld*, 2). With exegetical history, the aim is to clarify the views of historical figures as accurately and with as much objectivity as possible. In MFL Heidegger is more explicit about his opposition to this approach:

> [It] would be a totally misguided conception of the essence of philosophy were one to believe that one could finally distil *the* Kant, *the* Plato by cleverly calculating and balancing off all Kant interpretations or all Plato interpretations. This makes as little sense with Leibniz. What would result would be something dead 'Kant as he is in himself', Kant *an sich* ... The actuality of the historical, especially the past, does not emerge in the most complete account of the way it happened.
>
> (GA26 88/MFL 71–2)

Sleigh contrasts 'exegetical history' with 'philosophical history', the aim of which is to develop one's own philosophical views through a critical discussion of historical figures (ibid.). Exponents of the latter also aim to present accurate accounts of views of historical figures, but for the purposes of discussing ways in which they connect up with currently relevant philosophical issues.[4] In some cases, historical figures may turn out to have been 'right'. In many cases, however, they turn out to be 'wrong' and we 'learn' from their 'mistakes'. Another crucial difference between philosophical history and exegetical history, which Sleigh does not emphasize, is the likelihood that philosophical historians will be selective in the subject matter they discuss in a way that is influenced heavily by their own philosophical interests. By contrast the exegetical historian is more likely to be lead wherever the texts take her and/or to be interested in the overall philosophical worldview of her chosen philosopher. Is Heidegger doing philosophical history then? Whilst his approach is closer to philosophical than exegetical

---

[4]Sleigh puts his own work in the first of his two categories and a paradigmatic example of the latter for him is Jonathan Bennett. Bennett provides an interesting discussion of this method in the introduction to Bennett, *Learning from Six Philosophers*, 1–9. Something like this distinction can be found in Russell (*Critical Exposition of the Philosophy of Leibniz*, xii), in Rorty's ('Historiography of Philosophy') discussion of 'historical' and 'rational' approaches, and Laerke, Smith, and Schliesser's discussion of 'appropriationist' and 'contextualist' approaches (*Philosophy and Its History*, 1–3).

history, I think the answer is 'no', and that it is most obviously construed as a third way which is more readily associated with 'Continental philosophy'. I shall refer to this as 'dialogical history'.[5]

The contrast with Sleigh's philosophical history concerns the ways in which historians' philosophical interests enter into their interpretative practice. With philosophical historians, a key desideratum is the provision of a philosophically neutral interpretation that would be readily assented to by a chosen historical figure. Then, and only, is there an attempt to find out what might be of worth in that view. Consider now how Heidegger proceeds after offering his critique of exegetical history:

> The actuality of what has been resides in its possibility. The possibility becomes manifest as the answer to a living question that sets before itself a futural present in the sense of 'what can we do?' The objectivity of the historical resides in the inexhaustibility of possibilities, and not in the fixed rigidity of a result.
>
> (GA26 88/MFL 72)

Heidegger's understanding of being 'in dialogue with the thinkers of previous times' involves the idea that both sides of the conversation emerge only as a result of the questions asked. There is no sense in which one should articulate the views of another person on their own terms because this endeavour is futile. This is not to say that Heidegger abandons the thought that there is something beyond the text being revealed in the interpretation. However, he does not think that we can recover the definitive way in which the text did this originally, given that its capacity to reveal is partly a function of its place in a living dialogue.

Speaking of his analysis of Leibniz, Heidegger observes that 'We must suppose ... that this monadological interpretation of beings was initiated with an authentically philosophical intention' (GA26 94/MFL 76). Heidegger then represents himself as in pursuit of the authentic philosophical intention in question. This intention is revealed at the beginning of the discussion in *Pathmarks*, when Heidegger notes that he was 'guided by its perspective on the ecstatic being-in-the world of human beings granted by a look into the question of being' (GA9 373/P 63). In other words, Heidegger self-consciously reads Leibniz in such a way that he would stand in a productive relation to Heidegger's conception of the being of human beings. Indeed, for Heidegger it makes no sense to suggest that there could be anything else going on. He is not looking to discover Leibniz's views as an end in itself, but to enter into dialogue regarding a common subject matter, something that requires that the written text speak Heidegger's own language to at least some degree.

---

[5]My 'dialogical history' is essentially the unnamed third approach discussed in Laerke, Smith, and Schliesser (*Philosophy and Its History*, 3–4).

Whilst I think most of the material that I will discuss below should be regarded as the fruits of dialogical history, there is one place in MFL where Heidegger alludes to an even more radical approach:

> Our interpretation must risk proceeding beyond Leibniz, or, better, going back more originally to Leibniz – even with the danger of departing from what he in fact said.
>
> (GA26 88/MFL 72)

I shall call this fourth methodology 'creative history of philosophy'. Unlike dialogical history, it is history of philosophy that emerges from the reading of an historical text with no deep regard at all for the thoughts of the figure who is represented as the author. Indeed, at the limit, this approach requires nothing more than the attribution of a position to an historical figure, which is mediated by the reading of her or his writings. This method seems to me to be reminiscent of that adopted on at least one occasion by a well-known figure in the Anglophone tradition, namely P. F. Strawson. In *Individuals* (Strawson, *Individuals*), Strawson discusses Leibniz. But he does not claim to be trying to make sense of him on his own terms. Indeed, he admits that the name is being attached to a position that is a device of his own making.[6] Like Strawson, in this passage Heidegger appears sanguine about an issue that often comes up when people read his accounts of historical figures, namely, that one may be learning about Heidegger's views monologically rather than dialogically.[7]

2. SUBSTANCE AS MONAD

Bearing in mind these methodological considerations, I shall first turn to 'the guiding context of the problem' in MFL. Heidegger characterizes this as follows: 'On the basis of the monadology, we want to know about the being of beings' (GA26 89/MFL 72). As Heidegger notes in *Pathmarks*, Leibniz follows in a long tradition of substance metaphysicians. Thus, when discussing Leibniz, the question of the being of beings becomes a question about 'the substantiality of substances' (GA9 373/P63).

Heidegger begins his account by observing that the subject of his enquiry is 'the monadology' (ibid.). This might lead one to expect him to pay attention to the work that is commonly known as *The Monadology* (G VI, 607–23/ AG 213–5). However, this is not the case. Rather, Heidegger is indicating

---

[6] See Strawson, *Individuals*, Chap. 4.

[7] In fact, I see nothing in Heidegger's discussion of monads in MFL that should be regarded as creative history rather than dialogical history. However, Heidegger's 1951 lecture course on Leibniz, *The Principle of Reason* (GA 10), does appropriate Leibniz's texts in a much more radical way.

that he takes as his starting point Leibniz's use of the term 'monad' to pick out substances or genuine beings, and his thought that, for Leibniz, 'the essence of substance resides in it being a monad' (GA26 90/MFL 73; GA9 373–4/P 63). In MFL, Heidegger identifies the texts he regards as the most important for his purposes, and includes the *Discourse on Metaphysics* and correspondence with Arnauld. In fact, Heidegger's analysis makes no explicit appeal to these texts or any others written before 1694, and he mostly draws on the correspondence with De Volder of 1698–1706 and other texts dating from this time or later.[8]

The analysis begins with an explanation of Heidegger's understanding of 'monad'. Returning to the original Greek word μονας, he offers a number of candidates, 'the simple, the unity, the one ... the individual, the solitary' (GA26 89/MFL 72; GA9 373/P 63), and claims that all these are intended by Leibniz. Leibnizian substances have 'the character of the simple unity of the individual, of what stands by itself' (GA26 90/MFL 73; GA9 373/P 63). Immediately afterward, however, Heidegger provides a preview of where his elucidation of the substantiality of Leibniz's substances will lead. To be a monad will be to be that which 'simply and originally unifies and which individuates in advance' (ibid.). Here we begin to see that Heidegger thinks of the unity of monads as connected with the fact that substances *unify*, rather than their being uncomposed, and there is an indication that this involves 'looking ahead'.

At this point, *Pathmarks* skips five pages of the MFL text in which Heidegger contrasts Leibniz's account of substance with those of Descartes and Spinoza and explains why Leibniz rejects Descartes' attempt to 'see the being of physical nature ... in *extension*' (GA29 91/MFL 73). Here Heidegger mentions two important arguments against the Cartesian conception of body (see GA 26 93/MFL 75). The first trades on the fact that extension is essentially divisible cannot provide a principle of unity. The second, which appears in a letter to Bayle from 1687 (see GP III, 48), trades on two results which Leibniz established in his *Brief Demonstration*: 1) that Descartes's claim that the 'quantity of motion' (determined by the product of speed of a body and its size) in the universe as a whole is conserved is false; and 2) that what is actually conserved is a quantity measured by the product of the square of the speed of a body and its size.[9] Whilst the reasoning that Leibniz employs is somewhat unclear, his conclusion is not.[10] The conservation of the product of size and the square of speed gives us grounds to reject the idea that the substance of bodies can be extension alone. Instead, we must acknowledge that bodies are endowed with a distinct attribute force [*vis*], which is the cause of motion.

---

[8] See GA26 87/MFL 70–1.
[9] A VI, 4 2027-30/L 296–8.
[10] See Lodge, 'Force and the Nature of Body in *Discourse on Metaphysics* §§17–18'.

Heidegger draws these two strands together to 'adequately define monad' (GA9 374/P 64). Monads are 'not themselves in need of unification but rather that which gives unity' (GA26 95/MFL 77; GA9 374/P 64) and this unifying activity is connected with their force. They are said to be units that are 'primordially unifying' and characterized as 'primordially simple force' (ibid.). Here the term 'primordial' indicates more than a mere connection. Heidegger's thought is that 'unifying' and 'simple force' are terms that express different ways in which one and the same thing appears. The main task that Heidegger sets for himself in the remainder of his discussion is providing an explanation of this.

With primordial status conferred on the unity of monads and their being endowed with force, Heidegger turns his attention to the latter, observing that 'understanding the metaphysical meaning of the doctrine of monads depends on correctly understanding the concept of *vis primitiva*' (GA26 96/MFL 77; GA9 374/P 64) but also that it is crucial to his analysis that force be 'understood from the perspective of the problem of defining unity in a positive way' (GA9 374/P 64).[11] He begins his account of Leibniz's notion of force by turning to the journal article *On the Emendation of First Philosophy and the Notion of Substance*.[12] In Heidegger's eyes it is a crucial text. Thus, he tells his students, 'Whether we push through to the ontological significance of the monadology or remain stuck in the vapidity of popular philosophy depends on whether we understand this article or not' (GA26 96/MFL 78).

Heidegger draws attention to the way in which Leibniz, in this article, contrasts his views about force with those of the Scholastics through the distinction Leibniz draws between his own 'active force [*vis activa*]' and the 'mere power [*potentia nuda*]' (GP IV, 469/L 433). As Heidegger notes, Leibniz is keen to stress that, whereas the Scholastics conceived of active power as a faculty which gives rise to action only when there is an appropriate external condition, active force 'contains a certain acting that is already actual' (GA26 82/MFL 102; GA9 65/P 375). Leibniz allows that the activity of *vis activa* may require the 'removal of an impediment' (GP IV, 469/L 433), but he insists that 'some action always arises from it' (GP IV, 470/L 433). Thus, any impeding can only be partial. In order to make the difference between *vis activa* and the Scholastic notion of power clearer Heidegger observes:

> We call what Leibniz means here 'to tend toward ... ' or, better yet, in order to bring out the specific, already actual moment of activity: to press toward or *drive* [*Drang*]. Neither a disposition nor a process is meant, rather a letting

---

[11] In emphasizing this Heidegger's focus automatically moves away from many texts in which Leibniz himself starts by considering the material world and the role of force in the production of motion. For a survey of the ways in which Leibniz conceives of the relationship between the forces of bodies and his substance metaphysic, see Garber (*Leibniz: Body, Substance, Monad*, 2009, 287–301, 305–9).
[12] GP IV, 468–70/L 432–4.

something be taken on (namely, taken upon oneself), a being set on oneself (as in the idiom 'he is set on it'), a taking it on oneself.
(GA 26 102–3/MFL 82; GA9 375/P 65)

Here Heidegger decides not to render *vis activa* as 'force [*Kraft*]' which might 'suggest the idea of a static property' (GA 26 103/MFL 83; GA 9 376/P 65). Rather it is *Drang*, or 'drive' in the English translation by Heim.[13]

There are two crucial issues for Heidegger: First, he wants to emphasize the fact that, unlike Scholastic power which is 'merely ... a disposition which is about to act but does not yet act' (GA26 102/MFL 82; GA9 375/P 64–5), drive is 'self-propulsive' and, rather than being triggered by the presence of some external condition, 'leads into activity, not just occasionally but essentially' where 'this leading requires no prior external stimulus' (ibid.). The second issue is clarified in the revisions for *Pathmarks*, where scholastic power is said to be 'a present-at-hand capacity in something present at hand' (GA9 375/P 64–5). One feature of what it is for something to be 'present at hand' is that it is intelligible only from a third person perspective. As will become clear below, Heidegger takes Leibniz's conception of substance to be derived from the first person perspective.

At this point Heidegger reaches a provisional conclusion: For Leibniz 'every being has this character of drive and is defined in its being as having drive' (GA26 103/MFL 83; GA9 376/P 66). But nothing has been said to clarify 'the structure of drive' (ibid.), given that we have been offered no account what kind of activity it is that all beings engage in essentially. The remainder of Heidegger's discussion is largely concerned with this issue and its relation to the unity of monads.

3. THE EGO AS LEIBNIZ'S 'GUIDING CLUE' OR PARADIGMATIC IDEA OF BEING

Before tackling the question of the structure of drive, Heidegger observes that he 'needs to interpose another consideration' (GA26 105/MFL 85; GA9 380/P 68), one he famously takes up in the introduction to *Being and Time*.[14] If one wishes to clarify the being of beings one must find a way to make this being available for investigation. And, according to Heidegger, Leibniz, like Heidegger, does this by taking our own being as paradigmatic.

Heidegger offers a brief version of his own justification for doing this. On the one hand, we are beings who 'comport ourselves toward beings' (GA26

---

[13]Loemker uses the term 'conatus' here (L 433), which is commonly found in other translations of Leibniz.
[14]See SZ 26–7.

105/MFL 85; GA9 379/P 68),[15] something we do only because we have some understanding of the being of those beings. Whilst Heidegger does not think of this as essentially involving self-conscious awareness of the being of other beings – the absorbed playing of a piano would count, for example – in order for such directedness to be possible, that towards which there is comportment must be understood to be a being in some sense. One could not play the piano without 'knowing' that there was a piano there to be played. However, such comportments are exhibited by other people. So it is natural to ask why Heidegger thinks that progress requires that we focus on our own being. Here the thought is one that is central to Heidegger's approach in the late 1920s. Unlike the being of other beings our own being is always a 'concern for us' (GA26 106/MFL 85; GA9 379/P 68). And this is something which presupposes, like the playing of the piano, some kind of understanding of that being. Indeed, Heidegger claims that our understanding of the being of other beings is always conditioned by the way in which we understand our own being and that without such an understanding other beings would not show up for us at all.

There is no explicit ascription of this motivation to Leibniz. However, Heidegger seems to regard it as an explanation for the fact that Leibniz turns to the being of the 'I' in order to explicate the nature of monads. But more important for my discussion is Heidegger's observation that this is what Leibniz does, that is, that 'Constant regard for our own existence, for the ontological constitution and manner of one's own "I", provides Leibniz with the model of the unity that he attributes to every being' (GA 26 106/MFL 85/GA9 380/P 68).[16] Heidegger provides a number of passages which testify to this (see MFL 86–8/107–9; P 68–70/379–83), including the following, from a letter to Sophie Charlotte of 1702 'On What is Independent of Sense and Matter':

> The thought of myself, who perceives sensible objects, and the thought of the action of mine that results from it, adds something to the objects of the senses. To think of some color and to consider that one thinks of it are two very different thoughts, just as much as color itself differs from the 'I' who thinks of it. And since I conceive that other beings can also have the right to say 'I', or that it can be said for them, it is through this that I conceive what is called substance in general.
>
> (GP VI, 502/AG 188)

Although Heidegger thinks that Leibniz offers the ego as the guiding clue for our understanding of the being of beings, he is quick to draw us away from a 'superficial and arbitrary reading' that would lead us to think that

---

[15] I follow the translation from P which differs slightly from the version in MFL, but not in ways that are material to the present discussion.
[16] See note 19 below.

this 'is simply anthropomorphism, some universal animism by analogy with the "I"' (GA26 110/MFL 88; GA9 384/P 71). Instead he suggests we ask 'Which structures of our own Dasein are supposed to become relevant for the interpretation of the being of substance?' (GA26 110/MFL 88; GA9 384/P 72). And we are taken back to the questions: 'How does the drive that distinguishes substances as such confer unity? How must drive itself be defined?' (GA26 111/MFL 89; GA9 385/P 72).

4. HEIDEGGER'S ACCOUNT OF DRIVE

Heidegger moves next to the heart of his analysis. At this point things become hard to understand. However, it is possible to discern a number of central features, some of which provide ways of thinking about Leibniz's conception of monads that might be fruitful to exegetical or philosophical historians. I will try to clarify Heidegger's presentation in this section and return later to the things that I regard as potentially illuminating.

A key passage for Heidegger is from Leibniz's letter to De Volder of 30 June, 1704:

> It can be further suggested that this principle of activity [drive] is intelligible to us in the highest degree because it forms to some extent an analogue to what is intrinsic to ourselves, namely, representing and striving.[17]
>
> (GP II, 270)

In this passage, Leibniz draws attention to the intrinsic features of the ego. With these in focus, Heidegger's aim is to explain why Leibniz characterizes our activity thus, by appealing to the way in which drive might function as a unifying principle. It is Heidegger's view that this 'deepest metaphysical motive' was one that 'remained concealed from Leibniz himself' (GA26/ MFL 90; GA9 386/P 73). And, as a result of this Heidegger uses some neologisms to try to draw attention to phenomena that he thinks have hitherto received inadequate attention. The main elements in the position that Heidegger ascribes to Leibniz are as follows:

4.1. (a) Drive is simple

Since drive must confer unity, drive itself cannot be in need of unification. It is for this reason that drive (one might prefer to say, a being whose essence is drive) 'must itself be simple' and 'must have no parts in the sense of an aggregate, a collection ... must be an indivisible unity' (GA26 111/MFL 89; GA9 385/P 72).

---

[17]This is Heim's rendition of Leibniz. It includes Heidegger's interpolation of the term 'drive', but is not otherwise particularly idiosyncratic (see LDV 307).

4.2. (b) Drive unifies a manifold

Given that substances confer unity, there must be something that needs unifying. With drive playing the unifying role, it follows 'that there must be something manifold that [drive] unifies' (ibid.).

4.3. (c) The manifold is internal to drive

If drive is to unify a manifold there 'must be a manifold right there in the monad' and 'the monad as simple and unifying must as such predelineate the possible manifold' (GA26 111/MFL 89; GA9 385/P 72–3).

4.4. (d) The manifold has the character of drive

Given c), that is, that the manifold is in the monad which has drive as its essence, it follows that the manifold must 'have the character of drive, must have movement as such' (GA26 111/MFL 89; GA9 385/P 73), and that it 'is the changeable and that which changes' (ibid.).

4.5. (e) Drive is self-surpassing

From d) it follows that 'The manifold must have the characteristic of being driven for [*Gedrängte*]' (GA26 112/MFL 89; GA 9 385/P 73), or 'driven ahead [*Be-drängte*]' (ibid.); and 'There is thus in drive itself a self-surpassing; there is change, alteration, movement. This means that drive is what itself changes in driving on; drive is what is driven onward [*Ge-drängte*]' (GA26 112/MFL 89; GA 9 386/P 73).[18]

4.6. (f) Drive is prior to what is unified

Since drive is 'simply unifying' (GA26 112/MFL 90; GA9 386/P73), it must be 'an original organizing unification', and hence 'prior to that which is subject to unification' and 'anticipate by reaching ahead to something from which every manifold has already received its unity' (ibid.).[19] It is thus '*reaching out* [*ausgreifend*], and as reaching out, must be *gripping in advance* [*umgreifend*] in such a way that the entire manifold is already made manifold in the encircling reach' (ibid.).

Later Heidegger adds: 'it must already anticipate every possible multiplicity, must be able to deal with every multiplicity in its possibility ... Drive must therefore bear multiplicity in itself and allow it to be born in the driving' (GA26 114/MFL 91;GA9 387/P 74). In other words it is 'the

---

[18] I deviate here slightly from the translation at MFL 89 and the slightly revised translation at P 73.
[19] The translation follows P here.

source of multiplicity' (ibid.). Heidegger then equates this aspect of the monad with its being 'in its essence basically "re-presentative" [*re-präsentierend*]' (GA26 112/MFL 90; GA9 386/P 73).

4.7. (g) Drive, perception, and appetition

Soon after, Heidegger considers the relationship between the notion of representation that he has introduced and Leibniz's term 'perception [*perceptio*]' and the attendant notion of 'appetition [*appetitus*]' (GA 26 113/MFL 91; GA9 387/P 74). Here his view is that this bifurcation is due to the fact that Leibniz 'has not himself grasped the essence of *vis activa* with sufficient radicality ... In fact, drive is in itself already a perceptive striving or striving perception' (ibid.). However, Heidegger notes that 'appetition does not mean the same as drive' since '*Appetition* ... refers to a particular, essential, constitutive moment of drive, as does *perception*' (GA26 114/MFL 91;GA9 387/P 74).

4.8. (h) The nature of the manifold

It is not until the end of his account of drive that Heidegger clarifies what comprises the manifold, where we learn: 'What it unifies ... is nothing other than the transitions from prehension to pre-hension [*von Vorstellen zu Vor-stellen*]' (GA26 115/MFL 92; GA9 388/P75).[20]

Some of the forgoing account is relatively clear. From (a) it can be seen that drive, and hence that which has drive as its essence, is simple and has no parts. When combined with (d) and (e), we also see that this simplicity is supposed to be compatible with the fact that drive changes. The driving monad is to be construed as a 'continuously changing' unified being rather than a sequence of discrete, temporally successive, and qualitatively distinct states that stand in some unity-conferring relation to one another. The unifying function of drive is thus connected with the ways in which temporally removed stages of drive are nonetheless stages of the same drive. According to Heidegger, it is from the problem of the persistence of the monads despite their manifold differences that Leibniz's insistence on the unifying nature of the monad derives. The manifold that must be unified is, as we are told with (c) internal to drive, because it is, as we learn from (d), nothing other than drive itself, or more intelligibly, as we are told in (h), the successive moments in drive.

What is missing here is an account of how substances discharge this function. Here the material in (f) is pertinent. In some sense, as well as being the manifold, drive is prior to the manifold. The explanation of this finally

---

[20] Whilst 'Vorstellen' is more naturally translated as 'to imagine', Heim's introduction of the term 'prehension' (which means roughly the same as 'apprehension') is an effort to retain some of what he, plausibly, takes to be Heidegger's intention – namely, to indicate the fact that that which is unified is a temporal sequence of perceptual states.

emerges in the closing section of the lecture, where Heidegger's discussion turns to the role that drive plays in the individuation of monads. As we have seen, *qua* unifying, drive is prior to itself *qua* manifold. This is now cashed out in terms of a notion that is familiar to readers of Leibniz, namely a 'point of view'.

4.9. (i) Drive and points of view

Heidegger suggests that in driving perception 'there is a *possession of unity in advance* to which drive *looks*' (GA26 117/MFL 94; GA9 390–1/P 77), or 'there is a "point", as it were, on which attention is directed in advance' (GA 26 117/MFL 95; GA9 391/P 77), which is equated with 'the unity itself from which drive unifies' (ibid.). Furthermore, 'What is in advance apprehended in this viewpoint, is also that which regulates in advance the entire drive itself' (GA 26 117/MFL 95; GA9 391/P 78).

Heidegger's explanation of a point of view is not fleshed out. However, we can see that he thinks of drive as containing essentially a representative content to which all, and only, its stages contribute. Heidegger does not at this point bring in something else that Leibniz invokes at times, namely his analogy with laws of the series. But it seems to me that this is the kind of idea that he has in mind – i.e. that at every moment in its history a particular drive contains within it something that both encapsulates and unifies all the stages of its development and which finds its complete expression only through the entire development of the drive. If the monad were fully to develop there would be a vantage from which its history could be told that would include all its previous representations, and it is the latent presence of this vantage point throughout the development which enables the monad to proceed towards it. The monad drives towards a point from which the sum of its previous states could be represented, and hence unified, analogous to the way in which one might drive towards a destination, whose identity as a destination would be predicated on its including all the previous stages of the journey. And like the journeyer, the monad is only able to strive to get there because it already has some understanding of where it is going.

5. WHAT CAN BE LEARNED FROM HEIDEGGER'S ACCOUNT?

In the remainder of this paper I want to explore a number of ways in which Heidegger's reading of Leibniz might illuminate our thinking. I will in some cases embed this within the context of scholarly discussions. However, the extent to which I do this will be relatively limited. In particular, I will restrict myself to the ways in which Leibniz's thought has been discussed in Anglophone literature.

5.1. Textual points

I want to draw attention first to the texts that Heidegger discusses. As noted above, the earliest of Leibniz's writings is the 1694 article *On the Emendation of First Philosophy,* and many of Heidegger's quotes comes from the correspondence with De Volder and other works dating from the 1690s and early 1700s.

A positive feature of this is the fact that Heidegger emphasizes the significance of the 1694 publication. Whilst this text is well known to scholars, it is not the most easily accessible piece for Anglophone readers. It appears only in the rather dated (and expensive) volume edited by Loemker called *Philosophical Papers and Letters* and not the more recent (and cheaper) editions of Leibniz's works that students might expect to encounter.[21] Heidegger's suggestion that 'Whether we push through to the ontological significance of the monadology or remain stuck in the vapidity of popular philosophy depends on whether we understand this article or not' (GA26 96/MFL 78) may verge on the hyperbolic. However, in the article Leibniz sets out a manifesto regarding the way that metaphysics ought to be pursued as well as introducing the public for the first time to his view that the concept of force is crucial for a proper understanding of the concept of substance. Thus, I think Heidegger is right to suggest that this is a central text for our understanding of Leibniz's project. And, whilst well known to scholars, it is not currently accorded the status it deserves in the English-speaking world.

A less positive feature of Heidegger's choices is that they allow him to ignore many texts that would make it harder for him to present his account as if it were a dialogue with the essential core of Leibniz. Over the past thirty years or so Anglophone scholars have become particularly interested in the fact that Leibniz seems not have had a stable conception of substance over time, and, in particular, that it was only in the 1690s that he began to self-consciously articulate a view that contained the monadological theory.[22] And in his 2009 book *Leibniz: Body, Substance, Monad*, Daniel Garber provides a thoroughly documented account which would make it hard for anyone to maintain that Heidegger is correct to assume that the monadological metaphysics provides *the* single entry point for an analysis of Leibniz's understanding of substance.[23] The force of this criticism can be blunted by thinking about the way in which Heidegger's

---

[21] I have in mind AG and *Leibniz: Philosophical Writings* (Everyman's University Library, 1973) edited by G. H. R. Parkinson and translated by Mary Morris.

[22] For a summary of some of the main positions, see Lodge, 'Garber's Interpretations of Leibniz on Corporeal Substance in the 'Middle Years''. Other significant contributions include Baxter, 'Corporeal Substances and True Unities'; Phemister, 'Leibniz and Elements of Compound Bodies'; Hartz, *Leibniz's Final System*; Rutherford, 'Leibniz as Idealist'; Garber, *Leibniz: Body Substance, Monad*. This is not to say that the consideration of this idea is especially novel (see e.g. Cassirer, *Leibniz' System in seinen wissenschaftlichen Grundlagen*).

[23] See Garber (2009).

account of Leibniz is a function of the methodology that he employs. However, were one to approach Heidegger's account of Leibniz as if it were one from which one could discern a plausible account of *the* view of substance that Leibniz intended to articulate then one would be led astray.

5.2. Force as drive

Turning to the content of the account the Heidegger provides, there are a number of features, *qua* exegetical historian, that I find illuminating in his discussion of the nature of monads. The first is Heidegger's decision to translate *vis* as *Drang* or 'drive'. This choice usefully marks the fact that Leibniz's notion of force [*vis*] is a technical one whose connotation differs from what people are likely to associate with the term. *Vis* is not, as Heidegger points out, like the scholastic notion of power (a notion which has affinities with a dominant conception of dispositions that we find in contemporary metaphysics), in that it needs no external stimulus for its actualization. And, perhaps more importantly, it is unlike the notion of force that we are accustomed to from Newtonian physics which exists only where there is an interaction between distinct entities. Leibniz's *vis* is an intrinsic principle of motion.

5.3. Perception and appetite as founded in drive

A second point arises from Heidegger's analysis of perception and appetite. Here he claims that Leibniz's appeal to drive undercuts the need to appeal to the notion of appetite in understanding the grounds for intramonadic change. Heidegger's thought is that, whilst Leibniz did not grasp this, appetition, like perception, is a 'constitutive moment of drive' (GA 26 114/MFL 91;GA 9 387/P 74) and not something which gives rise to the changes that occur within monads. Heidegger does not clarify the way he is using the term 'moment' here. However, it seems that he is suggesting that perception and appetition are abstract ways of conceiving a principle of change that is essentially representational, or as Heidegger puts it that 'drive is in itself already a perceptive striving or striving perception' (ibid.).

Heidegger may be less generous to Leibniz than Leibniz deserves here. In the texts from the De Volder correspondence at the heart of Heidegger's discussion, there is nothing that invites the thought that perception and appetite are ontologically basic. It is the notion of *vis* that plays this role, or, as Leibniz prefers in a number of writings from the early late 1690s and 1700s the notion of 'dynamism' (LDV 12; 73; 241; 339), which he describes as 'an attribute from which change follows whose subject is substance itself' (LDV 73).[24] Nonetheless, the view that Heidegger presents is far from standard in secondary literature and seems to me to deserve further attention.

---

[24] Also see *On Body and Force Against the Cartesians* from 1702 (GP IV, 395/AG 252).

Whilst it may be that Leibniz regards perception and appetition as essential for an account of monadic change that answers to standard explanatory demands – e.g. that furnishes answers to determinate questions about what are regarded as particular changes in individual monads – this need not mean that he thinks the change is brought about through an interaction between two distinct faculties.

5.4. The ego as the guiding clue

Also prominent in Heidegger's analysis of monads is his claim that Leibniz's conception of the monad is drawn from our self-conception. This may seem obvious, given that Leibniz characterizes the intrinsic features of monads in quasi-psychological terms. However, it has been argued that the *New System* and texts from succeeding years are strikingly at odds with what we find around a decade earlier in this regard.

In *Leibniz: Body, Substance, Monad*, Daniel Garber offers a detailed analysis of Leibniz's conception of the matter and form of substances and their relation to the concept of force based on texts from the 1680s. Garber argues that, whilst Leibniz's account of the nature of minds is not radically different from the conception that he will later attribute to all monads, his account of the nature of material reality is. The material world is grounded in primitive active and passive forces which are 'not only the ground-level physical realities', but also 'the ultimate metaphysical realities that ground the created world' (*Leibniz: Body Substance, Monad*, 318). Garber's idea is that in the 1680s, Leibniz conceives of the nature of the material world as comprising the ground of dispositions to produce and resist changes in the locomotion of material things with no evidence of an attempt by Leibniz to understand the ground itself further.

Garber is at odds here with the prominent view of Robert Adams.[25] However, Adams's claim is not that Leibniz positively interprets the physical forces that Garber takes to be basic in monadological terms. Rather he is concerned that there is no 'Leibnizian' way to understand the dispositions to which Garber refers other than to identify them with the forces that will be given a 'psychological' reading in the later writings. In response Garber's main complaint is that Adams, like others before him, is illegitimately projecting a later conception back into these texts.[26]

Heidegger draws attention to texts in which Leibniz explicitly claims that we are to understand the nature of *vis* by appeal to our internal awareness of the operations of our souls. It is clear in these passages that Leibniz thought that subjecting the world experienced in sense to a certain kind of intelligibility was necessary for a proper understanding of its nature. The texts Heidegger highlights remind us that Adams's philosophical intuition about how the

---

[25] See Adams, *Leibniz: Determinist, Theist, Idealist*, 327–8, 338, 347–9.
[26] See Garber (2009, 166–72).

forces of bodies might be understood is not merely an interpreter's imposition. Leibniz raised the issue himself and chose to approach it by appeal to self-knowledge.[27] Furthermore, the epistemic issues that Leibniz emphasizes in the passages Heidegger highlights provide us with clues as to what kinds of textual evidence we might look for to help decide whether Garber's position is defensible.

5.5. Unity and the unifying function of drive

Finally, I think there are things to be learned from Heidegger's emphasis on monads being understood (at best latently by Leibniz) as unities in the sense that they play a *unifying* role. Here Heidegger goes against a reading of why Leibniz regards substances as unities that is clearly found in texts spanning his career. In these passages Leibniz takes it as given that there are composite, and hence divisible, entities, and argues from the convertibility of unity and being that the existence of these composites requires that there be entities that are not divisible from which they are composed.[28]

By contrast, when Heidegger considers the thesis that monads are unities, he is adamant that Leibniz's commitment to this stems from the need for beings that *unify*, something that he thinks could not be done by entities that were not themselves unities. It should not be thought that the need for monads (or perhaps substantial forms in the 1680s) to play a unifying role is ignored by commentators. Indeed, to the extent that people take Leibniz's commitment to corporeal substances seriously (i.e. extended beings which are nonetheless unities), they include a commitment to entities that perform such a function. But this is not where Heidegger places his focus. As we saw above, the manifold that Heidegger regards as unified by the monad is internal to the monad itself; the heart of Leibniz's authentic concern when he focuses on unity (although perhaps unnoticed by Leibniz himself) is connected with the persistence of individual monads.

One might balk at the thought that it is this function that is the ultimate basis for Leibniz's obsession with monadic unity. But as far back as 1977, Robert Sleigh Jr. raised the issue of whether monads should be thought to have temporal parts and, if so, whether this might compromise their reality as beings which have identity over time.[29] Sleigh himself did not offer a solution to the problem in his paper, and, as far as I am aware, it is a challenge that has not received much attention and about which there is much more to

---

[27] Also, see Phemister, *Leibniz and the Natural World*, 177–82 for discussion of the first- and third-person perspective in Leibniz.

[28] This strategy is common in the correspondence with Arnauld and other documents from the 1680s (see Garber, 'Leibniz and the Foundations of Physics' and 2009 chapter 2). But Leibniz also argues for the existence of monads in something like the same way in his correspondence with De Volder in the early 1700s (see LDV 275; 285–7; 301–3) and as late 1714 in the *Monadology* (GP VI, 607/AG 213).

[29] See Sleigh, 'Leibniz on the Simplicity of Substance', 120.

be said. One recent effort attempts to deal with the issue by placing monads outside of time and treating the attribution of temporal parts to them as ideal.[30] This solution has a lot of merit, but it requires deflating passages in which Leibniz appears to understand the existence of monads in such a way that their temporality involves a robust notion of succession.[31] Whilst I do not have the space to examine it further here, the account I have extracted from MFL seems to me to do more justice to this and to provide a starting place for further investigation.

5.6. Methodological considerations

The last place in which I think readers of Leibniz may have something to learn from Heidegger is by considering the methodology that he employs. I began my discussion by distinguishing four ways in which one might read historical texts: via exegetical history, philosophical history, dialogical history, or creative history. The key contrast between the first two conceptions and the second two is that with the former there is a sense in which understanding what the historical figure intended is a regulative ideal. With dialogical and creative history, there is no such pretension. The text is engaged with, but the enquiry is entirely one of first order philosophy. It is the truth, or at least the philosophical interest, of what the interpreter takes away from the reading that is of importance and, if there is little left that the author might have attested to, that does not undermines the results.

One important challenge that Heidegger poses comes from a key presupposition he brings to bear on reading texts, namely that, *qua* interpretations, they will always have a lot of the interpreter in them. Leibniz viewed from nowhere may serve as a regulative hermeneutic ideal for some, but as Heidegger insists, this ideal can never be reached. And his recommendation is that we get as clear as possible about what we are asking of a text when we engage with it, and what it is about our interests that leads us to ask our questions.

One might think this point need no longer be emphasized given its prevalence in twentieth century hermeneutics. But if one turns to articles and books written about Leibniz in English today, the majority read as if the author lacked any self-conscious engagement with these issues. Thus, Heidegger's challenge remains a significant one and he offers an important reminder that providing an account of Leibniz *an sich* can at best be a regulative ideal. Nonetheless, there is still a difference between Heidegger's method and reading texts such that one tries to allow the author to 'speak', and/or where one extends this to listening carefully to the way in which the text interacts with contemporary and antecedent philosophical

---

[30]See Whipple, 'Continual Creation and Finite Substance in Leibniz's Metaphysics'.
[31]For example, see LDV 289.

voices. The claim that we cannot remove our interests entirely as readers is clearly not the same as the claim that we cannot stand at different removes from them.

A second issue that Heidegger forces us to think about is whether it is legitimate to recoil, as many historians of philosophy seem to, from the idea of dialogical or creative history of philosophy. It seems to me that this urge to recoil might come from three worries, each of which meets opposition from Heidegger's practice.

A first worry might be that such a cavalier approach to text is likely to be of no help in understanding what a cherished dead figure actually thought. But, of course, there is nothing in the dialogical or creative method that precludes the generation of interpretations that might be appropriated by the exegetical or philosophical historian. Indeed, in considering ways in which Heidegger's account in MFL and P might illuminate Leibniz, my approach has been to point to things that have just this kind of status. The evidence here suggests that there is enough continuity between the Leibniz that emerges and what currently passes for the concerns of Leibniz 'himself' that our understanding of the latter might be augmented by reading Heidegger. And it seems obvious this is could also be true of a radically creative historian of philosopher. If clear about what they are doing, historians of any stripe are free to cherry pick for their own ends.

A second possible worry is that the dialogical and creative historians are likely to end up being neither good historians nor good philosophers. But this kind of reaction seems to amount to little more than the claim that they will not be doing history of philosophy as the critic conceives it, or philosophy as she conceives it. In the case of Leibniz, I have suggested that Heidegger may have things to say that the community of exegetical historians and philosophical historians could usefully appropriate, but it would be no serious criticism of his endeavour if he did not. Nor is it the case that the dialogical and creative historians could reasonably be accused of failing to do good philosophy *qua* dialogical and creative historians. Dialogical and creative history as I am conceiving of them are not first order philosophy. Rather they are engagement with historical texts in the service of generating first order philosophy.

Presumably the only legitimate complaint here could be that dialogical and creative historians do not generate good philosophy as their end product. This worry will be something that applies, or not, *ad personam*. And it would be hard to deny that my canonical dialogical and creative historians have enriched philosophy in ways that would have been impossible without this approach. Absent his dialogical and creative engagements with key figures in the history of philosophy there simply is no Heidegger, and without *Individuals* it is hard to believe there would have been the Strawson we know either. And if Heidegger and Strawson are not to one's taste, then it should be easy to see elements of this kind of approach in the work of other people once one starts to look. To pick two radically different examples, Giles Deleuze's treatment of Leibniz in *The Fold* (Deleuze, *The Fold*) and Jerry Fodor's *Hume Variations* (Fodor, *Hume Variations*) come to my mind, and I would have thought that anyone should be able to find dialogical and creative history that fits their first order philosophical tastes. If there is a residual concern, I suspect that it may be due to irritation that dialogical and creative history does not require engagement with the hard won interpretations of those more concerned with exegesis. But if so, there is a danger that this may be due to ressentiment as much as anything else.

A final worry might be harder to get people to admit explicitly, namely, that if one admits that dialogical and creative history have value, one might cede too much to those who have little sympathy for the role of history of philosophy as a branch of academic research. This is a natural worry, given the experiences that some historians of philosophy have had. The feeling of being marginalized and of being regarded as less philosophically able than one's peers, is something that I think many carry with them in the professional environment that is twentieth first century Anglophone academic philosophy.

People know of the famous sign that was posted on the door of Princeton philosopher Gilbert Harman in the 1980s: 'History of Philosophy: Just Say No!' In explaining his actions, Harman acknowledged that history of philosophy could reasonably be conducted only as proposed by exegetical historians. More precisely, he observed:

> [A] study of the history of philosophy tends not to be useful to students of philosophy. (Note 'tends'.) Similarly, it is not particularly helpful to students of physics, chemistry, or biology to study the history of physics, chemistry, or biology.[32]

In other words, he rejected the legitimacy of philosophical history, dialogical history, and creative history of philosophy. Harman claimed he had acted in good faith. But unfortunately his error was typical of the way those in positions of power dismiss those with opposing views. Even allowing that it was a careless act, it was still an act which carried the implication that the efforts of people identifying themselves as trying to do history other than exegetically were worthless. Furthermore, it implied that even the acceptable version of the history of philosophy was likely to be of no value to those studying the subject itself.

The story of Harman's sign did the rounds within the profession as a joke. But it is difficult to escape the thought that it was an expression of views that could have reasonably been taken as posing an existential threat to those working in academia as historians of philosophy in philosophy departments. Furthermore, the story seemed to capture a more general sense of how history of philosophy was regarded in many Anglophone philosophy

---

[32]See http://philosophy.princeton.edu/about/eighties-snapshot.

departments at the time. It is not hard to see why some historians might have felt like members of a threatened group who needed to develop a respectable identity within the space that was being offered to them. Indeed, with the retrenchment in place and feelings passed on to their academic children, it would be surprising if there were not those, perhaps subliminally, who spurn dialogical and creative philosophy partly because they are worried that those who stray beyond the exegetical are likely to put the future of the history of philosophy as a distinct academic specialization in jeopardy.

But it is thirty years since Harman and, to whatever extent these ghosts still haunt, it is time to move on. Harman was just wrong. Whilst he himself may have been incapable of enriching his philosophy by turning to the history of the subject, there is no sense in which dialogical history, creative history, or indeed the philosophical history that Harman also had in his sights, sit uncomfortably with first order philosophizing. Furthermore, there is no reason to think that rich exegetical history is any different, provided its relationship to philosophical history, dialogical history, and creative history is cultivated in a certain way. It is time to reverse whatever normative shifts Harman's diatribe might have engendered.

If we are to reap the huge philosophical benefits that are there to be reaped from a subject whose 2000-year-plus history bequeaths us numerous texts written by highly sophisticated philosophers, far from consigning them to history, we should regarded them as documents that can enrich our philosophical lives every time we visit them with open eyes.[33] This might look like an advertisement for privileging the other forms of history of philosophy over exegetical history. However, it seems better to me to recognize exegetical history as something that can play a vital role in allowing the philosophical historian to do her work. By consulting works of exegetical history her readings are likely to be enriched, whether through fine-grained exegesis of her chosen author or consideration of the broader philosophical context that can only arise from careful consideration of such things as the roles of 'minor' figures whom she might find less philosophically inspiring. There is an inevitable division of labour here given the amount of potentially fruitful data, and whilst this may mean that exegetical history is sometimes perceived as a kind of under-labouring, there is clearly no shame to be had in that, especially where the labour requires talents that many of the consumers do not themselves have.

But where does this leave dialogical and creative history? The dialogical and creative historians I have focused on are people who seem to have

---

[33] In saying this I do not, of course, mean to imply there are no other reasons to value the study of the history of philosophy. Furthermore, as Sarah Hutton has emphasised recently (Hutton, '-Blue-Eyed Philosophers Born on Wednesdays'), there are important connections to be explored between engaging in careful exegetical history of philosophy and the recovery of the writings of women philosophers and other neglected figures. However, even in this case, my inclination is to think that the greatest value of this kind of project lies in the role that it can play in shaping philosophy itself.

engaged with texts in isolation from other readers. And with such people in mind, it might appear that this practice has little to gain from the results of other modes of engaging with historical texts. Here I think it has to be admitted that dialogical and creative history may well be able to operate successfully in isolation, provided that they are practised by philosophers who have questions of their own that grip the philosophical community. But it seems at least possible that benefits might arise from interaction with philosophical and exegetical history for any would-be creative historians today.

In conclusion then, my reflections on Heidegger's approach to reading suggest a way in which we might articulate the virtues of a methodological pluralism in the history of philosophy that accords significant status to work that is at the more exegetical end of the spectrum. But at the same time, there is an important presupposition. The justification that I have offered for this is, in the end, that it serves the greater goal of advancing the progress of philosophical thinking itself. It is immensely enjoyable to do exegetical history of philosophy, and it fosters a kind of intellectual community that is, at its best, a common enterprise in which there is a great deal of mutual respect. But the argument I am advancing is predicated on the thought that it can, and perhaps should, gain its true meaning from its service to philosophy itself. And it goes without saying that all philosophers are under-labourers in that respect.

BIBLIOGRAPHY

ABBREVIATIONS
A = *Gottfried Wilhelm Leibniz: Sämtliche Schriften und Briefe*, edited by Deutsche Akademie der Wissenschaften (Darmstadt and Berlin: Akademie, 1923-). Cited by series, volume, and page.
AG = *Leibniz: Philosophical Essays*, edited and translated by Roger Ariew and Daniel Garber (Indianapolis: Hackett, 1989).
GA9 = *Martin Heidegger, Wegmarken* (Frankfurt am Main: Klosterman, 1996).
GA10 = *Martin Heidegger, Der Satz vom Grund.* Translated as *The Principle of Reason*, translated by R. Lilly (Bloomington: Indiana University Press, 1991).
GA26 = *Martin Heidegger, Metaphysische Anfangsgründe der Logik* (Frankfurt am Main: Klosterman, 1978).
GP = *Die Philosophische Schriften von Gottfried Wilhelm Leibniz*, 7 vols., edited by C. I. Gerhardt (Berlin: Weidmann, 1875–90; reprint ed. Hildesheim: Olms, 1960). Cited by volume and page.
L = *Gottfried Wilhelm Leibniz, Philosophical Papers and Letters*, 2nd ed., edited and translated by Leroy Loemker (Dordrecht: D. Reidel, 1969).

LDV = G. W. Leibniz, *The Leibniz-De Volder Correspondence*, edited and translated by Paul Lodge (New Haven, CT and London: Yale University Press, 2013).

MFL = *Martin Heidegger, The Metaphysical Foundations of Logic*, trans. Michael Heim (Bloomington: Indiana University Press, 1984).

P = *Martin Heidegger, Pathmarks*, edited and translated by William McNeill (Cambridge: Cambridge University Press, 1998).

SZ = *Martin Heidegger, Sein und Zeit* (Tübingen: Niemeyer, 1957).

Adams, Robert M. *Leibniz: Determinist, Theist, Idealist*. Oxford: Oxford University Press, 1994.

Baxter, Donald L. M. 'Corporeal Substances and True Unities'. *Studia Leibnitiana* XXVII, no. 2 (1995): 154–84.

Bennett, Jonathan. *Learning from Six Philosophers: Descartes, Spinoza, Leibniz, Locke, Berkeley, Hume*. Vol. 1. Oxford: Oxford University Press, 2001.

Cassirer, Ernst. *Leibniz' System in seinen wissenschaftlichen Grundlagen*. Marburg, Germany: N. G. Elwert'sche, 1902.

Deleuze, Giles. *The Fold: Leibniz and the Baroque*. Translated by Tom Conley. Minneapolis: University of Minnesota Press, 1992.

Fodor, Jerry. *Hume Variations*. Oxford: The Clarendon Press, 2003.

Garber, Daniel. 'Leibniz and the Foundations of Physics: The Middle Years'. In *The Natural Philosophy of Leibniz*, edited by K. Okruhlik and J. R. Brown, 27–130. Dordrecht, The Netherlands: Reidel, 1985.

Garber, Daniel. *Leibniz: Body, Substance, Monad*. Oxford: Oxford University Press, 2009.

Hartz, Glenn. *Leibniz's Final System: Monads, Matter, and Animals*. London: Routledge, 2006.

Hutton, Sarah. ''Blue-Eyed Philosophers Born on Wednesdays': An Essay on Women and History of Philosophy'. *The Monist* 98, no. 1 (2015): 7–20.

Laerke, Mogens, Justin Smith, and Eric Schliesser. *Philosophy and Its History: Aims and Methods in the Study of Early Modern Philosophy*. Oxford: Oxford University Press, 2013.

Lodge, Paul. 'Force and the Nature of Body in *Discourse on Metaphysics* §§17–18'. *Leibniz Society Review* 7 (1997): 116–24.

Lodge, Paul. 'Garber's Interpretations of Leibniz on Corporeal Substance in the 'Middle Years''. *The Leibniz Review* 15 (2005): 1–26.

Phemister, Pauline. 'Leibniz and Elements of Compound Bodies'. *British Journal for the History of Philosophy* 7, no. 1 (1999): 57–78.

Phemister, Pauline. *Leibniz and the Natural World*. Dordrecht, The Netherlands: Springer, 2005.

Rorty, Richard. 'The Historiography of Philosophy: Four Genres'. In *Philosophy and Its History*, edited by R. Rorty, J. B. Schneewind, and Q. Skinner, 49–76. Cambridge: Cambridge University Press, 1984.

Russell, Bertrand. *A Critical Exposition of the Philosophy of Leibniz*. Cambridge: Cambridge University Press, 1900.

Rutherford, Donald. 'Leibniz as Idealist'. *Oxford Studies in Early Modern Philosophy* 4 (2008): 141–90.

Sleigh, Robert C., Jr. 'Leibniz on the Simplicity of Substance'. *Rice University Studies* 63 (1977): 107–21.
Sleigh, Robert C., Jr. *Leibniz and Arnauld: A Commentary on Their Correspondence*. New Haven, CT: Yale University Press, 1990.
Strawson, P. F. *Individuals*. London: Methuen, 1959.
Whipple, John. 'Continual Creation and Finite Substance in Leibniz's Metaphysics'. *Journal of Philosophical Research* 36 (2011): 1–30.

ARTICLE

FIVE FIGURES OF FOLDING: DELEUZE ON LEIBNIZ'S MONADOLOGICAL METAPHYSICS

Mogens Lærke

> This article is about Gilles Deleuze's book *Le Pli. Leibniz et le Baroque* from 1988. It shows how Deleuze's notion of folding captures some basic intuitions in Leibniz and how they relate to each other. To this purpose, I propose five figures, all referring to the same basic fold, all illustrating how the consideration of such figures allows developing central elements of Leibniz's monadology. These figures can help, I hope, alleviate some of the fundamental difficulties in understanding Deleuze's approach to the Monadology from the non-Deleuzian perspective of contemporary Leibniz scholarship and give a sense of the synthetic, explanatory force that Deleuze's notion of folding has in relation to Leibniz's monadological metaphysics.

1. INTRODUCTION

Gilles Deleuze's book *Le Pli. Leibniz et le Baroque* from 1988 has been used for a great many purposes.[1] It is much appreciated by literary scholars, art academy students and architects who have adopted, or rather adapted, many of Deleuze's ideas about 'the fold to their own practices, often widening the spectrum of applications but equally often losing sight of the systematic, conceptual background of Deleuze's concept and how exactly it relates to Leibniz's monadological metaphysics. While early Deleuze scholars who studied *Le Pli* often produced mediocre commentary,[2] more recent

---

[1] I will use the following abbreviations for Leibniz's work: A = *Sämtliche Schriften und Briefe*; AG = *Philosophical Essays*; Arthur = *The Labyrinth of the Continuum*; GM = *Leibnizens mathematische Schriften*; GP = *Die philosophischen Schriften*; L = *Philosophical Papers and Letters*.

[2] For example, the translator of *Le Pli*, Tom Conley, writes in his introduction regarding Leibniz's logic: 'By means of Leibniz's innovation [...] the subject is enveloped in the predicate, just as Proust's intension is folded into its effect' (Conley, 'Translator's Foreword', xiv). In Leibniz, of course, it is the other way round, i.e. the predicate that is included in the subject, and Deleuze never says otherwise. Comments like this display deep ignorance of basic features of Leibniz's metaphysics, a curious comfortableness with nonsensical

commentators, including Simon Duffy, Daniel W. Smith and others, have contributed substantially to our understanding of how *Le Pli* relates to other work by Deleuze.[3] Deleuze scholars are however rarely sufficiently equipped with plain historical knowledge about Leibniz's philosophy to assess the merits of Deleuze's book as a piece of Leibniz scholarship. As for Leibniz scholars, the situation is the opposite. In most cases unequipped with broader knowledge of Deleuze's philosophy, they struggle to make good sense of *Le Pli*'s densely written analyses which draw on Deleuze's other work in multiple ways.[4] This said, use of *Le Pli* among non-Deleuzian Leibniz scholars is not infrequent. We can, for example, find friendly references in Christia Mercer's *Leibniz's Metaphysics* from 2001 and active use of important aspects of the Deleuzian perspective in Richard Arthur's recent book *Leibniz* from 2014 (see Mercer, *Leibniz's Metaphysics*, 8; Arthur, *Leibniz*, chapters 3, 5, and 9). In none of these cases, however, does the use of Deleuze's book presuppose any deeper understanding of the notion of folding and how it is related to the basic structure of Leibniz's monadological metaphysics.

Although I do think Deleuze's analysis of Leibniz in *Le Pli* has considerable appeal, the following is not a plea in favour of his reading, but only an attempt to show how the notion of folding can capture, in an extraordinarily synthetic way, some basic intuitions in Leibniz and how they relate to each other. To this purpose, I provide five simple graphic figures that I hope can help in understanding what Deleuze had in mind. These figures have their limitations. They do not capture *all* the intuitions behind Deleuze's fold-concept, neither do they comprehend the full scope of the intuitions that they *do* capture. Most importantly, in order not to complicate matters unduly, I everywhere represent the fold in two dimensions as a folded *line*. There is, however, a good argument to be made for rather representing it as a folded three-dimensional *surface*. Moreover, the figures below do not relate the *dynamic* aspect of the fold, that is, the fact that it moves, folds in and folds out. This is an aspect that becomes relevant, for example, in Leibniz's account of birth and death, according to which 'what we call generation are developments and growths, as what we call deaths are enfoldings and diminutions' (Leibniz, *La Monadologie*, § 73, GP VI, 619, trans. AG, 222). A perfectly adequate description of the folding paradigm would thus not be in terms of a curve, or folded line, but rather in terms of a moving, folded surface, maybe something like the surface of the sea or a bed sheet on a washing line. I have nonetheless opted for considering the properties

---

statements, and warrant great caution when using the translation. Throughout this article, I refer only to the original text and provide my own translations.

[3] Among other contributions, many of which are referenced below, see in particular the volume edited by S. Van Tuinen and N. McDonnell: *Deleuze and* The Fold*: A Critical Reader*.

[4] I should here make an exception of the splendid article by Bouquiaux, 'Plis et enveloppements chez Leibniz', 39–56.

of a simple curve in order not to get entangled in mathematical considerations too complicated for my main purpose, which is more pedagogical than philosophical (let alone mathematical). Despite such limitations, however, I hope and believe that the five figures of folding I develop below can help provide a better grasp of Deleuze's interpretation.

2. MONADOLOGICAL METAPHYSICS

Before delving into the details of Deleuze's reading, some preliminary remarks should be made about how he approaches Leibniz's metaphysics in *Le Pli* and how this reading relates to the *Monadology*.

Let me begin by forestalling a possible misunderstanding. *Le Pli* is not a book about Leibniz. It is first of all a work in the history of aesthetics that should be classified along with Deleuze's other work on aesthetics from the 1980es, notably the short book on Francis Bacon and the two volumes on cinema.[5] Indeed, *Le Pli* is no more a work about Leibniz than *L'Image-temps* and *L'Image-movement* are works about Henri Bergson or Charles Sanders Peirce. Leibniz's philosophy is, in *Le Pli*, used to give conceptual consistency to the otherwise very elusive epochal and aesthetic category of the Baroque. Like other commentators, most importantly Walter Benjamin, Herbert Knecht and Peter Fenves,[6] Deleuze believes that baroque aesthetics finds a philosophical equivalent in Leibniz's monadology. Hence, for Deleuze, Leibniz 'gives [the Baroque] the philosophy it was lacking'.[7] However, *Le Pli* also provides the basic elements of a global interpretation of Leibniz's metaphysics and it is the way that those elements are organized around the notion of folding that is the topic of the following. But it is important to keep in mind that this is not the primary purpose of Deleuze's book.

It should equally be kept in mind that there are other books where Deleuze engages with Leibniz's philosophy.[8] Throughout *Spinoza et le problème de l'expression*, particularly in the conclusion, Deleuze proposes very dense comparative analyses of Spinoza and Leibniz centred around their common anti-Cartesianism and different concepts of expression (see Deleuze, *Spinoza et le problème de l'expression*, 299–311). In *Différence et répétition*, when criticizing the so-called 'regimes of representation', he describes Leibniz as the philosopher who, like Hegel albeit in a different

---

[5]See Deleuze, *Logique de la sensation*; *L'Image-mouvement*; *L'Image-temps*. On this point, see also Peden, *Spinoza Contra Phenomenology*, 305, n. 10.
[6]See Benjamin, *The Origin of German Tragic Drama*; Knecht, *La Logique chez Leibniz*; Fenves, '*Autonomasia*', 432–52. On Leibniz and Benjamin in Deleuze, see Lærke, 'Four Things Deleuze Learned from Leibniz', 25–45.
[7]Deleuze, *Le Pli*, 173.
[8]For discussion of the development of Deleuze's reading of Leibniz, see Robinson, 'Events of Difference', 141–64; and Smith, 'Deleuze on Leibniz', 127–8.

way, made representation infinite, making him one of the major villains in Deleuze's catalogue of philosophers, occupying the spot right after the German master of negativity (see Deleuze, *Différence et répétition*, 62–3, 71, 119, 338–40). In a more affirmative mode, Deleuze's conceptualization of the 'alogical incompatibility' between ideal series that he terms 'vicediction', and that constitutes a very important element in his account of how individual things come to be, is basically modelled on Leibniz's notion of 'compossibility' (see Deleuze, *Différence et répétition*, 68–9, 338–9; see also *Le Pli*, 79). This also comes through in *Logique du sens*, sixteenth series, where Deleuze appeals explicitly to the notion of 'compossibility' in order to conceptualize what he calls the 'static ontological genesis' (Deleuze, *Logique du sens*, 133–41, 198–203). In these earlier texts, however, Deleuze's use of Leibniz is surgical. His aim is not so much systematic exposition as it is to highlight specific features of Leibniz's philosophy, sometimes only for comparative purposes. We find a more comprehensive reading of Leibniz's philosophy if we turn to Deleuze's lectures at the University of Paris VIII in 1980 and in 1986–7, most of which are available online in French transcription and English translation.[9] Those lectures, eminently pedagogical and fairly straightforward in their approach to Leibniz's texts, are immensely helpful for grasping Deleuze's reading in the *Le Pli*, otherwise a difficult work to penetrate for the uninitiated.

None of Deleuze's works, including *Le Pli*, provide a full exposition of Leibniz's philosophy. If, however, we combine the early readings and the Paris VIII lectures with the insights developed in *Le Pli*, it is possible to extract a near-complete reading of Leibniz's monadological metaphysics from Deleuze's texts. I here speak very consciously of a 'monadological metaphysics', although it is an expression that Leibniz never uses, that Deleuze does not explicitly use, and that I would myself use only with the most extreme caution. Anyone who has witnessed the 'genetic' turn in Leibniz studies over the last decades – spearheaded by Michel Fichant and Daniel Garber, and followed by a great many other Leibniz scholars (see Fichant, 'L'invention métaphysique', 7–140; Garber, *Leibniz. Body, Substance, Monad*) – would immediately object: What is 'monadological metaphysics'? When and where exactly does Leibniz develop it? Is it even possible to determine exactly when it is complete? Does it not constantly evolve? Multiple problems regarding the real, historical elaboration of Leibniz's doctrine will arise.

---

[9]See http://www.webdeleuze.com. The site includes lectures from 15/04/1980, 22/04/1980, 29/04/1980, 06/05/1980, 20/05/1980, 16/12/1986, 201/01/1987, 27/01/1987, 24/02/1987, 10/03/1987, 17/03/1987, 18/03/1987, 07/04/1987, 12/05/1987, 19/05/1987/, 20/05/1987, 25/05/1987, and one undated lecture from early 1987. Deleuze's lectures from 28/10/1986, 4/11/1986, 18/11/1986, 06/01/1987, 13/01/1987, 27/01/1987, 03/03/1987, 17/03/1987, 28/04/1987, 05/05/1987, 19/05/1987, 02/06/1987 are also available as sound-files on *Gallica*, the digital library of the *Bibliothèque Nationale de France* (http://gallica.bnf.fr).

Deleuze is perfectly indifferent to such concerns. Whether he speaks of the 1673 *Confessio philosophi* or the 1710 *Essais de théodicée*, the 1676 *Pacidius philalethi* or the 1714 *Principes de la nature et de la grâce* – it is all for Deleuze the same doctrine in different formulations. Unsurprisingly, the commentator for whom he declares most unequivocally his admiration is Michel Serres.[10] In his 1968 thesis *Le Système de Leibniz et ses modèles mathématiques*, Serres develops a structuralistic reading of Leibniz according to which everything Leibniz ever wrote, from the *De arte combinatoria* to the *Monadology*, represents different aspects of one and the same system, 'scenographic' variations on the same virtual 'ichnography' as Serres paraphrases a distinction we find in some late Leibniz texts.[11] Hence, for Serres, as also for Deleuze who follows him, the 1714 *Monadology* is just a highly compressed summary of a 'monadological metaphysics' that permeates all of Leibniz's philosophical writings.

This stubbornly synchronic view of Leibniz's philosophical enterprise is, from the historical point of view, an obvious flaw in Serres's approach and it is so in Deleuze's as well. Fortunately, in Deleuze's case, this flaw is in part counterbalanced by another flaw, namely the fact that Deleuze, in contrast to Serres, mostly ignores the parts of Leibniz's philosophical work that fall outside the standard corpus contained in the editions of Gerhardt and Couturat. In particular, he shuns the philosophical work of the young Leibniz, most of which is published in the series VI of the Academy edition, an edition that Deleuze simply does not use.[12] Consequently – with a few exceptions such as the *Confessio philosophi* and the *Pacidius philalethi* –, he rarely cites texts that do not belong to Leibniz's 'mature' philosophical work, that is to say, from around or after 1686, making it considerably less shocking, although still not quite pardonable, to talk about Leibniz's philosophy as a single 'system' or 'monadological metaphysics'.

3. THE FOLD IN LEIBNIZ'S TEXTS

What does this 'monadological metaphysics' look like for Deleuze? As is often the case in Deleuze's highly synthetic interpretations of other

---

[10] Deleuze repeatedly endorses Serres's readings (see *Le Pli*, 12, n. 18; 25, n. 9; 30, n. 16; 40, n. 5; 102, n. 38; 104, n. 2; 120, n. 13; 176, n. 17). He does, however, disagree with Serres on a particular point concerning the position of a mathematical law in relation to the series it grounds: Is the law outside or inside the series? Deleuze, contrary to Serres, argues that it is outside, but stresses that 'this is one of the only points where I cannot follow Serres' (*Le pli*, 68, n. 20).

[11] See, for example, Leibniz to Des Bosses, 15 February 1712, in *The Leibniz-Des Bosses Correspondence*, 232-3.

[12] In his defense, it should be mentioned that volume A VI, iv (covering the years 1677-1690) only appeared in 1999. The other three volumes were available to him: A VI, i (1663-1672) appeared in 1930, A VI, ii (1663-1672) in 1966, and A VI, iii (1673-1676) in 1980.

philosophers, his reading of Leibniz is governed by a single interpretive move, expressed by a single operative notion. That notion is, unsurprisingly, that of the *fold*. As Deleuze concludes his book, 'we remain Leibnizian because it is all still about folding, unfolding, refolding' (Deleuze, *Le Pli*, 189). Let me in this section say a few words about the textual basis for that general interpretive strategy.

Deleuze's references to texts in Leibniz where there is explicitly a question of folding are very few. The most important one is to a passage of a 1676 paper, the *Pacidius philalethi*, where Leibniz explains how 'the division of the continuum must not be considered to be like the division of sand in grains, but like that of a sheet of paper or tunic into folds. And so, although there occurs some folds smaller than others infinite in number, a body is never thereby dissolved into points or minima' (A VI, iii, 555, trans. Arthur, 185).[13] This early passage is Leibniz's most developed development of the paradigm of folding. The second text that Deleuze references (but does not quote) is the *Protogaea,* Leibniz's account of the origin of the earth, written around 1690–1, and in particular chapter VIII which is concerned with 'deposits of metals in the earth and a description of veins' (see Leibniz, *Protogaea*, chap. VIII, 20–5). Finally, Deleuze references a letter to Des Billettes from 1696, where, in a context similar to that of the *Protogaea*, Leibniz is trying to explain the origin of magnetism in the following hermetic passage, describing the geological formation of veins of magnetic material:

> [...] it seems that the notion that particles are transformed into folds is not necessary; it suffices that after a while, the passages accommodate themselves to what passes through them that going backwards is somehow prevented and, so to speak, like brushing against the sense of the hair (GP VII, 453).

On these textual grounds alone, Deleuze's approach is clearly a philological non-starter. The first text, the *Pacidius philalethi*, supports Deleuze's use of the image of the fold, but this is a very early text that can hardly be qualified as part of his monadological metaphysics – in fact, in this text Leibniz embraces a metaphysical position closer to occasionalism.[14] As for the other, later, passages, they hardly advance Deleuze's case, since the *Protogaea* only suggests folds and the letter to Billettes declares them unnecessary rather than the contrary. In the end, Deleuze's most promising references are some suggestive passages in the *Monadology*, § 61, where Leibniz writes that 'a soul can read in itself only what is distinctly represented; it cannot unfold all its folds at once, because they go to infinity' (GP VI,

---

[13]Deleuze cites the text in *Le Pli*, 9

[14]See the exchange on 'transcreation' in *Pacidius philalethi*, October 1676, A VI, iii, 567–9, trans. Arthur, 21–7. Leibniz here holds that things do not pass from one state to another in virtue of an inherent power of change, but that change should rather be explained by the fact that 'a permanent substance [...] has both destroyed the first state and produced the new one'. On this text, see also Levey, 'The Interval of Motion', 371–416.

617, trans. AG, 221), and in the *Principes de la nature et de la grâce*. § 13, where Leibniz states that 'one could know the beauty of the universe in each soul, if one could unfold all its folds, which only open perceptibly with time' (GP VI, 604, trans. AG, 211).

If, however, we turn to texts more recently made available in the Academy edition, and in particular to the massive volume VI, iv, published in 1999, we find further material bolstering Deleuze's case in texts that most likely were perfectly unknown to Deleuze himself. Thus, in a fragment entitled *Definitiones cogitationesque metaphysicae*, written around 1678–81, Leibniz notes that 'the parts of one body constitute one continuum. For a unity always lasts as long as it can without destroying multiplicity, and this happens if bodies are understood to be folded rather than divided' (A VI, iv, 1401, trans. Arthur, 249–51). And again, in the *Conspectus libelli elementorum physicae* from around 1678–9: 'It is seen that in truth everything is fluid, and is only folded in different ways, without the continuity ever being broken' (A VI, iv, 1900). And finally, in an undated text on Saint Augustin, Leibniz notes: 'The whole universe is one continuous body. Nowhere divided, but transfigured like wax, or folded in different ways like a garment' (A VI, iv, 1687). The image of folding thus shows up fairly frequently in texts from the late 1670s and early 1680s, but remains a component in later texts as well. For the strict historian of philosophy, Deleuze's recourse to the image or notion of folding thus presents itself as textually somewhat better grounded than one would suspect at first sight, indeed better than what Deleuze probably himself suspected.

4. THE FOLD OF THE WORLD

Let me now turn to the graphic figures. According to Deleuze, the world is for Leibniz comparable to an infinitely folded curve that extends to infinity.[15] Figure 1 represents a segment of that infinite fold or curve, with two stationary points, P and Q, and an inflexion point Z.

Such a curve can be expressed mathematically by means of a function. Now, for Leibniz, there is a specific function for any possible curve passing through any possible configuration of points, and, conversely, any possible configuration of points can be generated by a specific function. Leibniz writes:

> Thus, let us assume, for example, that someone jots down a number of points at random on a piece of paper, as do those who practice the ridiculous art of geomancy. I maintain that it is possible to find a geometric line whose notion

---

[15]Deleuze, Lecture on Leibniz, 06/05/1980: 'But you can make this abstraction, you consider the world. How do you consider it? You consider it as a complex curve [...]. For Leibniz, that is what the world is'. See also *Le Pli*, 81; and *Pourparlers*, 217.

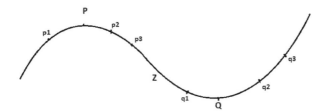

Figure 1. The fold of the world.

is constant and uniform, following a certain rule, such that this line passes through all the points in the same order in which the hand jotted them down. And if someone traced a continuous line which is sometimes straight, sometimes circular, and sometimes of another nature, it is possible to find a notion, or rule, or equation common to all the points on this line, in virtue of which these changes must occur (A VI, iv, 1538, trans. AG, 39).

In this passage, drawn from the Section 6 of the *Discours de métaphysique*, Leibniz explains that any compossible set of things or events – that is, things or events that can exist together without contradiction – can constitute a world, that is to say, be brought under a given function governing that world as its universal law. The underlying idea here is that any possible world unfolds following a pre-established plan in the way that a given mathematical curve is drawn in accordance with a given function, but that some worlds develop according to a simpler function than others.[16] God will choose to bring into existence the world that balances the simplicity of the function to the richness in variations of the resulting curve. God, Leibniz continues, will create the world 'which is at the same time the simplest in hypotheses and richest in phenomena, as might be a geometric line whose construction would be easy but whose properties and effects would be very remarkable and of a wide reach' (A VI, iv, 1538, trans. AG, 39).

Moreover, according to Leibniz, 'each possible world depends on certain principal designs or purposes of God which are distinctive of it'.[17] Hence, each possible world is characterized by a certain number of 'remarkable' events, supported by infinitely many 'ordinary' events leading up to each of them and connecting them to each other (before Caesar crosses the Rubicon, he first approaches the Rubicon, gets off his horse, wades a few steps into the water, etc. ...). We can observe that on the curve: P and Q, the two 'remarkable' points, are related by any number of 'ordinary' points in between: $p1$, $p2$, $p3$, and $q1$, $q2$, $q3$, etc.[18] Elaborating on an

---

[16] On this passage, see also Lærke, 'Compossibility, Compatibility, Congruity'.

[17] Leibniz to Arnauld, 14 July 1686, A II, ii, 73.

[18] See Deleuze, *Le Pli*, 81, 121. Deleuze also distinguishes between the 'remarkable' and the 'ordinary' in terms of the 'regular' and the 'singular'. Hence, 'a singularity is surrounded by a cloud of ordinaries or singulars' (*Le Pli*, 81).

example that Deleuze occasionally uses, one could, for example, argue that our world is built up around the two remarkable or singular events in world history that are Adam's fall and the expiation of our inherited sin through the death and resurrection of Christ.[19] Everything that happens in between these 'principal designs', P and Q, happens in order to establish continuity between them. The inflexion point Z is where the one is prolonged into the other, the place where the ordinary events begin to relate to and support the second rather than to the first remarkable event. In this case, the inflexion point Z could, for example, mark the birth of Christ or the place in the complete biblical story where we pass from the Old to the New Testament. In sum, the world we live in is the best, brought into existence by God because it could be created according to simple laws giving rise to maximal variety, and because it included a number of particularly 'remarkable' events or 'principal designs' that God felt it was best to include. God then created the world in such a way that those features were connected to each other in a world history that is maximally rich in phenomena while still following regular, simple laws.

5. THE FOLDS OF MATTER

Concretely, in the existing world, that is to say, the phenomenal physical world, the fold of the world is expressed as a 'fold of matter'. So, a rock, a plant, or a human body, are all constituted by folded matter. As we have already seen, there are numerous texts where Leibniz uses the notion of folding in order to account for the apparently paradoxical double affirmation that matter is, at the same time, continuous and actually divided to infinity. That matter is actually infinitely divided is a very characteristic feature of Leibniz's conception of the physical world. It runs counter to standard Aristotelian wisdom according to which actual infinity implies contradiction.[20] Hence, according to the *Monadology*, § 65, 'each portion of matter is not only divisible to infinity, as the ancients have recognized, but is actually subdivided without end, each part divided into parts having some motion of their own' (GP VI, 618, trans. AG, 221; see also GP VI, 599, trans. AG, 207). At the same time, however, such division cannot be division into an infinite number of discrete parts, since Leibniz is also committed to a principle of continuity, that is, the principle according to which nature is

---

[19] See Deleuze, Lecture on Leibniz, 22/04/1980: 'Why is Adam's sin included in the world that has the maximum of continuity? We have to believe that Adam's sin is a formidable connection, that it is a connection that assures continuities of series. There is a direct connection between Adam's sin and the Incarnation and the Redemption by Christ. There is continuity.'
[20] For Aristotle, see *Metaphysics*, II, 2, 994a, in *The Complete Works*, vol. II, 1570.

'full' (GP VI, 617, trans. AG, 221) and 'makes no leaps'.[21] This brings us to the second figure of the fold, a kind of microscopy of the basic fold, illustrating how it can be subdivided into infinitely many other folds (Figure 2).

The image of a 'fold going to infinity' or infinitely folded curve very effectively expresses a conception of the extended world that honours those two criteria at the same time: infinity and continuity. Moreover, it allows grasping an important point in Leibniz's anti-atomistic conception of the physical world, namely the fact that no physical body as such allows for a precise limitation, but that the border between the inside and the outside of the body is blurred and fluffy, exactly because any closer look at a given body will reveal that its surface is infinitely folded. Leibniz writes to Arnauld:

> [...] the very shape which is of the essence of a delimited extended mass is never really exact and determined in nature, because of the actual infinite division of the parts of matter. There is never a globe that is not uneven and no straight line without curves mingled into it, neither is there any curve of a certain finite nature which is not mixed with some other [curve], and this in the small parts as much as in the large ones, something which implies that shape, far from being constitutive of bodies, is not just an entirely real and determined quality outside the mind, and one can never assign to a body a certain precise surface as one could if there were atoms.[22]

In relation to these aspects of Leibniz's theory – the continuity and infinite division of matter, and the fluffy shape of extended things – the idea of the 'folds of matter' has both intuitive appeal and good support in the texts. Moreover, the 'mingling' of curves into curves that the letter to Arnauld speaks of is quite nicely illustrated by Figure 2. So far, the folding paradigm captures in a very straightforward way Leibniz's basic conceptions.

6. EVENTS

From here on it gets a little more complicated. For the folds of matter cannot be folds of *passive* matter. Leibniz was one of the foremost critics of the Cartesian physics, in particular of the Cartesian conception of extension as essentially inert mass. In his 'reformed physics', or 'dynamics', Leibniz thus formulated a physics according to which physical objects are not determined in the standard mechanist way by extension and movement, but by

---

[21] See, for example, Leibniz, *Nouveaux essais sur l'entendement humain*, IV, xvi, § 12, A VI, vi, 473. For a somewhat dated, but very insightful paper on the principle of continuity, see Arthur, 'Leibniz in Continuity', 107–15.
[22] Leibniz to Arnauld, 9 October 1687, A II, ii, 250. See also Leibniz to Arnauld, 16 April 1687, A II, ii, 171.

Figure 2. The infinite division of matter.

action and force. As Leibniz put it in the *Discourse of Metaphysics*, 'the nature of body does not consist merely in extension, that is, in size, shape and motion' but 'force [...] is something more real' (A VI, iv, 1545, trans. AG, 44, and A VI, iv, 1559, trans. AG, 51). Hence, for Leibniz, the physical world of bodies is animated by derivative forces that are metaphysically grounded in the primitive forces of individual substances, or monads. Each physical body must be considered a specific configuration of such derivative forces. Correlatively, on the metaphysical level, a body is an aggregate of substances endowed with primitive active and passive force.[23]

But how are we to reconcile these two apparently different conceptions of extended bodies and the physical world, that is, as *folded matter* and as complexes of *derivative forces*? Deleuze clearly tries to reconcile them, affirming, for example, that 'matter is a force that refolds itself incessantly'.[24] But how exactly is this supposed to work? We need here to make some very basic mathematical considerations. Let us look again at the fold. A fold is an inflexion on a curve, that is, a curve folds insofar as it curves. Now, as everyone knows, Leibniz is the inventor of the differential calculus. We all also know that the calculus is a kind mathematics that is concerned with the infinitely small and, in some interpretations, with limit values. But what sort of operation is it exactly that the differential calculus allows us to perform when it comes to curves? Installing our fold in installing in a coordinate system coordinate system can help elucidating that question (Figure 3).

One of the things that the differential calculus allows doing is, in each point of a curve, to determine a particular value that is habitually called the 'derivative' or the 'differential quotient', expressed by the formula dy/dx. The differential quotient dy/dx is the limit of the ratio between Δx and Δy when Δx is vanishing, or infinitely close to 0. We express that as follows:

$$\frac{dy}{dx} = \lim_{\Delta x \to 0} \frac{\Delta y}{\Delta x}$$

---

[23]For synthetic introduction to Leibniz's dynamics, see, for example, the *Specimen dynamicum*, 1695, GM, 234–54, trans. AG, 117–38.
[24]Deleuze, Lecture on Leibniz, 16/12/1986.

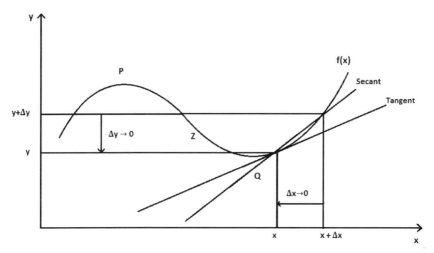

Figure 3. The differential calculus.

What does this value dy/dx express in relation to the curve? It does not express the position of the curve, which is given by the simple co-ordinates (x, y). Neither does it express the relation between two points on the curve, for example, the relation between (x, y) and $(x+\Delta x, y+\Delta y)$. This relation is expressed by the secant passing through those two points. Instead, since Δx is vanishing, and Δy along with it, dy/dx expresses the *inclination* of the curve, not between two points, but *at* a point, that is to say, the inclination of the tangent passing through the point (x, y). This tangent indicates exactly what *happens* in (x, y), that is, not where or what the curve is, but what the curve *does* at this particular point. Moreover, it indicates how the curve in any given point is already moving on to something else, how it is already in the process of taking up a relation to another point. Hence, the calculus allows grasping the nature of the curve, not as a composition of *points* (x, y), but as continuous succession of differential relations, each expressing the *curving* itself, that is, the way in which the curve *varies*, or the fact that something *happens* in each point of the curve (see Deleuze, *Le Pli*, 24; see also *Pourparlers*, 214). In short, the differential calculus allows the mathematician to express a curve as a continuous series of *events*. This is the reason why, already in *Logique du sens*, Deleuze affirms that 'the first theoretician of the event is Leibniz' (Deleuze, *Logique du sens*, 200; see also *Le Pli*, 25, and *Pourparlers*, 216).

Considering how the differential calculus describes curves thus serves to make mathematical sense of the notion that the existing world is not a set of objects, but rather a series of events.[25] According to this paradigm, all things

---

[25]To be sure, there is an element of anachronism in my explanation to the extent that Leibniz himself did not interpret the derivative dy/dx in terms of a limiting process. It is, however, a

are fundamentally conceived as something that *occurs*, eliminating the difference between things and events: 'Sinning', 'Crossing the Rubicon', or 'the Pyramids' are all happenings, as it were, complex spatiotemporal organizations of events, each expressing their particular segment of the world fold.[26]

7. INCLUSION IN THE MONADS

As we have seen, according to Deleuze, the physical world is like an infinitely long sequence of inflexions on the world fold, a long series of events. This world of matter consists, as Leibniz puts it, of 'real phenomena'.[27] They have reality insofar as they are in fact enacted. Before those events are realized in the existing world, they are however first constituted as what Deleuze calls 'ideal events'. As he writes in *Logique du sens*: 'What is an ideal event? It is a singularity. Or rather it is a set of singularities, singular points which characterize a mathematical curve [...]. They are points of turning back, of inflexion, etc.' (Deleuze, *Logique du sens*, 67). Why does Deleuze speak here of inflexions in terms of 'ideal' events and how do such ideal events come into existence as 'real phenomena'?

According to Leibniz, the world is *pre-established*. It is conceived by God before any of the individual substances that inhabit it are created. Hence, before they are realized, all events of the world have an *ideal* or merely *possible* existence in God's mind.[28] The same applies to all conceivable worlds:

---

graphically clear way of illustrating how Deleuze understood the value dy/dx as a kind of pure relation, a relation with no *relata*, and how this fits with his conception of the Leibnizian predicate as pure event. It is worth noting that, in his interpretation of the calculus, Deleuze owed a great deal to the structuralist reading proposed by the Nicolas Bourbaki group. On this, see Deleuze, 'A Quoi reconnaît-on le structuralisme?', 299–335. For the notion of structure according to the Bourbaki mathematicians, see, for example, 'L'architechture des mathématiques', 35–47. On Deleuze's interpretation of Leibniz's *calculus*, see also Duffy, 'Deleuze, Leibniz and Projective Geometry in *The Fold*', 129–47.

[26]Deleuze, *Le Pli*, 103. By this should not be understood that such organizations are *constituted within* a given spatiotemporal framework, but rather that they are *constitutive of* the spatiotemporal framework, to the extent that, for Leibniz, space and time are nothing by orders of simultaneity and succession derived from the relative order among phenomena (or here: events). See, for example, Leibniz, Third paper to Clarke, 25 February 1716, GP VII, 363–4, trans. AG, 325.

[27]For the notion of real phenomena, see Leibniz, *De modo distinguendi phaenomena realia ab imaginariis*, approx. 1683–85, in A VI, iv, 1498–504, trans. L, 363–6.

[28]I here follow Deleuze's (in my own view problematic) reconstruction which describes the 'existentification' of things as a realization of the possible that is distinct from the actualization of the virtual. Deleuze thus argues that there is 'actuality that remains possible and which is not necessarily real' (see *Le Pli*, 140). There is however in Leibniz's own texts some evidence in favour of the contrary notion that, by virtue of being conceived in God's mind, the merely possible rather remains real while not being actual. The problem may simply be one of terminology.

they exist *qua* possible in God's mind. In Deleuze's reading of Leibniz, such possible worlds are, in a sense, worlds without subjects, in any case not constituted by individual substances.[29] These worlds and the individuals that inhabit them exist only in God's mind, as sequences or bundles of ideal events. For example, there is in God a concept of Adam's sinning, indeed concepts of every property or event that pertains to Adam, together constituting the complete concept of Adam. But Adam himself, an individual substance expressing those properties or events, is as yet nowhere in view: 'The world, as the common term expressed by all the monads, is pre-existent in relation to its expressions [...]. God has not created Adam the sinner, but first the world in which Adam sins' (Deleuze, *Différence et répétition*, 68). Individual substances, or monads, only become relevant when God decides to create one of the possible worlds, namely the one he considers to be the best. For, in order for ideal events to pass from a state of mere ideality or possibility into reality, they must be *enacted by substances*. Events cannot realize themselves. In this context, Leibniz often appeals to the scholastic axiom according to which actions are actions of subjects (*actiones sunt supppositorum*).[30] For example, in order for the ideal event of sinning to become real, it must be enacted by Adam; or in order for the ideal event of crossing the Rubicon to become real, it must be enacted by Julius Caesar. The way God brings a sequence of ideal events, that is, a possible world, into existence is thus by creating the subjects or monads that incarnate that sequence of events: God creates Adam so that he can sin and Caesar so that he can cross the Rubicon. Strictly speaking, then, God does not create the best possible *world*. Rather, he creates the *substances* appropriate for realizing or incarnating that world. One text where Leibniz intimates such a reading of his doctrine is the section 14 of the *Discours de métaphysique*:

> For God, so to speak, turns on all sides and in all ways the general system of phenomena which he finds it good to produce in order to manifest his glory, and he views all the faces of the world in all ways possible, since there is no relation that escapes his omniscience. The result of each view of the universe, as seen in a certain position, is a substance which expresses the universe in conformity with this view, should God see fit to render his thought actual and produce this substance.
>
> (A VI, iv, 1549–50, trans. AG, 46–7)

---

[29]This is one crucial point where I think Deleuze's reading of Leibniz's modal ontology diverges importantly from current standard readings according to which the complete concepts of possible individuals conceived in God's mind can, in some important sense, also be understood as possible substances. On Deleuze's reading, and rigorously speaking, there is no such thing as a possible substance. Complete concepts of things are, insofar as they are conceived in the divine mind, more like bundles of predicates. It would take us too far afield to assess the merits of either reading here.
[30]See for exemple Leibniz, *Discours de métaphysique*, sect. 8, A VI, iv, 1539–40. For a commentary, see Fichant, 'Actiones sunt suppositorum', 135–48.

Deleuze formulates this relation between the ideal events of the world and the active powers of created substances by means of the Leibnizian notion of *inclusion*. Inclusion is closely related to the logical relation of inherence, associated with Leibniz's famous principle of *inesse*, that is, the principle developed in the *Discours de métaphysique* according to which, in all true propositions, the predicate is included in the subject (A VI, iv, 1539–42, trans. AG, 40–2). Logically speaking, monads are like subjects.[31] The world, or the events of the world, are like properties or predicates that inhere in those subjects.[32] Correlatively, the substances enact ideal events or actualize the world by *including* them, that is to say, by *expressing* them internally.[33] Hence, the events of the chosen possible world become real phenomena in virtue of being expressed as properties of created substances. Now, monads are soul-like substances, and their activity is an inner one that can be modelled on our mental life.[34] Their internal activity – basing ourselves on the analogy with the soul – can be formulated in terms of perception and appetition (i.e. the tendency to move from one perception to the next) (Leibniz, *La Monadologie*, § 14–15, GP, 608–9, trans. AG, 214–5). Hence, the way in which monads realize or enact events is by *perceiving* them or *representing* them.[35] Ideal events, in short, become real phenomena by being projected into the perceptive universe of individual substances.

We can illustrate this intricate relation between world and monads by means of the following fold figure (Figure 4).

The world-fold is included in m(P) from the perspective of P and in m(Q) from the perspective of Q in such a way that m(P) and m(Q) *express* the fold from the viewpoint of P and Q, respectively. This expressive activity of monads consists in perception or mental representation. What God does,

---

[31] Deleuze implicitly relies on Louis Couturat's classic conception of the monad as 'nothing but the logical subject elevated to substancehood' (see Couturat, 'Sur la métaphysique de Leibniz', 9).

[32] This implies that all properties expressed by predicates are, when properly analysed, reducible to events or actions. Thus, according to Deleuze, the fundamental propositional form in Leibniz's logic is not the traditional attributive proposition *S is P* (such as 'Adam is a sinner'), but rather a verbal proposition of the form *subject–verb* ('Adam sins') (see *Le Pli*, 69–72; *Pourparlers*, 218; *Logique du sens*, 200–1; Lecture on Leibniz, 20/01/1987). Leibniz's favourite examples do indeed suggest something like that: 'Adam sins', 'Caesar crosses the Rubicon', 'Alexander combats Porus and Darius.' The point is that the predicates ascribed to subjects in propositional logic correspond to actions ascribed to subjects or, more precisely, to events enacted by subjects. The reading is far from being unwarranted (for a similar reading of the Leibnizian predicate, see Fichant, 'L'invention philosophique', 57).

[33] Leibniz mainly develops his notion of expression in the *Discours de métaphysique* and the correspondence with Arnauld. The literature on the topic is extensive. See, for example, Kulstad, 'Leibniz's Conception of Expression', 56–76; and Swoyer, 'Leibnizian Expression', 65–99.

[34] See, for example, Leibniz to Arnauld, 28 November (8 December) 1686, A II, ii, 121.

[35] A monad is 'by its very nature representative' (*La Monadologie*, § 60, GP VI, 617, trans. AG, 220).

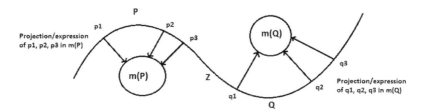

Figure 4. Inclusion.

then, when creating the world is that he 'lodges' a monad in each and every fold of the world, and the way in which they actualize the world is by expressing or perceiving that world from the perspective of the particular fold they occupy. The inflexions of the fold are, as it were, projected into – or included in – the monads m(P) and m(Q). We find these monads even in the minutest folds of the world. All the way down, we will find yet more folds with yet more monads lodged in them, each perceiving the world from their particular place. This is why Leibniz affirms that 'there is world of creatures, of living beings, of animals, of entelechies, of souls in the least part of matter' (Leibniz, *La Monadologie*, § 66, GP VI, 618, trans. AG, 222).

8. PERSPECTIVISM

It is important to realize how each individual substance, on account of the 'connection of things' as Leibniz sometimes puts it (Leibniz, *Discours de métaphysique*, sect. 8, A VI, iv, 1541, trans. AG, 41), represents not only a part of the world, but all of it:

> [...] since everything is connected because of the plenitude of the world, and since each body acts on every other body, more or less, in proportion to its distance, and is itself affected by the other through reaction, it follows that each monad is a living mirror [...], which represents the universe from its own point of view [...].
> (Leibniz, *Principes de la nature et de la grâce*, § 3, GP VI, 599, trans. AG, 207)

Hence, our monads m(P) and m(Q) do not only express the events related to P and Q, respectively, but both monads express or perceive the same complete world of events. They do however not express it from the same viewpoint, since m(P) expresses it from the position P and m(Q) from the position Q. Hence, according to the *Système nouveau*, 'every substance represents the whole universe exactly and in its own way, from a certain point of view, and makes the perceptions or expressions of external things occur in the soul at a given time [...]' (Leibniz, *Système nouveau*, 1695, GP IV, 484, trans. AG, 143.). In the *Monadology*, § 57, Leibniz puts the point as follows:

> Just as the same city viewed from different directions appears entirely different and, as it were, multiplied perspectively, in just the same way it happens that, because of the infinite multitude of simple substances, there are, as it were, just as many different universes, which are, nevertheless, only perspectives on a single one, corresponding to different points of view of each monad.
> (GP VI, 616, trans. AG, 221)

We here touch upon what Deleuze designates as Leibniz's 'perspectivism'. Such perspectivism, Deleuze insists, involves no relativism or subject-dependence of the perception of the world, but quite to the contrary a dependence of the subject itself on the viewpoint from where it is determined to perceive the world:

> We call it [i.e. the subject] a viewpoint to the extent that it represents a variation or an inflection. Thus is the foundation of perspectivism. It does not mean some dependence in relation to some given defined subject. On the contrary, the subject is that which comes to the viewpoint, or rather which inhabits the viewpoint.[36]

The subject does not ground a viewpoint on the world, but 'the subject is that which comes to a viewpoint',[37] exactly because the fold of the world, that which determines the order of viewpoints, is pre-established. Hence, in order to become real, the world must be included, or projected into the monads' respective perceptive universes, but the way in which the world is perceived by the monads is not determined by the monads themselves, but by the particular position they occupy in relation to the world they perceive, or express. The formula that sums up this expressive relation between the world and the subjects that inhabit it is thus the following: The world exists *in* the monads, but the monads exist *for* the world (Deleuze, *Le Pli*, 141–2). Deleuze often repeats this formula in order to stress the non-relativist point that the world is logically prior to the subjects that come to inhabit it: 'There is antecedence [of the world] in relation to the monads, even though a world does not exist outside the monads which express it [...]' (Deleuze, *Le Pli*, 81).

9. CLEAR ZONES OF EXPRESSION

Each monad perceives and expresses the entire world. However, due to their different situations in relation to the fold, that is, the fact that they are lodged in different folds, each monad does not express the universe *in the same way*: '[...] every substance expresses the entire universe, and [...] its individual

---

[36] Deleuze, *Le Pli*, 27.
[37] Lecture on Leibniz, 16/12/1986.

essence consists in nothing but this expression of the universe taken in a certain sense'.[38] The individuality of each monad is thus determined, not by *what* it represents – because all monads represent the same world –, but by the *manner* or *way* in which it represents it. It is for this reason that Deleuze affirms that Leibniz's philosophy is a *mannerism*, as opposed to an *essentialism* (Deleuze, *Le Pli*, 27, 55, 70–2, 76).

This brings me to my fifth and final figure, which concerns Leibniz's theory of clear and obscure perception. The fact that each monad perceives the world from its particular point of view entails that there are differences in clarity of their perception of the world, some segments being expressed clearly by some monads and only obscurely by others. Hence, Leibniz writes in the *Monadology*, 'a soul can read in itself only what is distinctly represented; it cannot unfold all its folds at once, because they go to infinity' (GP VI, 617, trans. AG, 221). Each monad has its specific 'clear zone of expression': '[...] in every created monad there is only a small part that is distinctly expressed [...] and all the rest, which is infinite, is only expressed confusedly' (Leibniz, *Eclaircissement des difficultés que Monsieur Bayle a trouvées*, GP VI, 553). This is illustrated by Figure 5.

In the *Principes de la nature et de la grâce*, § 2, Leibniz writes that each monad includes the world as multiplicity in a unity, just like 'in a center or point, though being entirely simple, we find an infinity of angles formed by the lines that meet there' (GP VI, 599, trans. AG, 207). On the figure, then, in order to determine the clear zone of expression corresponding to a given monad, we should, in each point on the curve, draw the lines perpendicular to the tangents of the curve, the so-called 'normal' lines (e.g. the normal lines in p1, p2, p3, and q1, q2, and q3). Those normal lines that converge and meet in the position, or *situs*, of a given monad determine the segment of the world fold that that monad will express clearly; those normal lines that do not converge towards the *situs* of the monad, on the contrary, will determine the part of the world fold that it will express only obscurely. In our case, the normal lines for the segment pertaining to the stationary point P, that is, the lines in p1, p2 and p3, all converge towards m(P). After the inflexion-point Z, in q1, q2, and q3, on the contrary, they diverge with regard to m(P), but converge with regard to m(Q). This illustrates the fact that m(P) is in charge of expressing clearly the fold constructed around the point P up until the inflexion point Z, where monad m(Q) takes over, being in charge of expressing clearly the fold around Q.

This difference between clear and obscure zones of expression can also be expressed in terms of Leibniz's famous theory of minute perceptions, mostly clearly laid out in the preface to the *Nouveaux essais sur l'entendement humain*. Elaborating on Locke's analysis of the feeling of 'uneasiness' in the *Essay concerning human understanding*, Leibniz here stresses the importance of this feeling as something that marks out psychologically our

---

[38]Leibniz to Arnauld, 30 April 1687, A II, ii, 167.

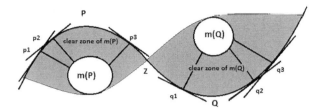

Figure 5. Individuation and clear zones of expression.

confused perception of the entire universe (A VI, vi, 38–68). Leibniz alludes to the theory in the *Principes de la nature et de la grâce* in the following terms: 'Each soul knows the infinite – knows all – but confusedly [...]. Confused perceptions are the result of impressions that the whole universe makes upon us' (GP VI, 604, trans. AG, 211). According to this theory, each clear perception is a macro-perceptual integration of infinitely many minute perceptions:

> Since each distinct perception of the soul includes an infinity of confused perceptions which embrace the whole universe, the soul itself knows the things I perceived only so far as it has distinct and heightened [*relévées*] perceptions; and it has perfection the extent that it has distinct perceptions. Each soul knows the infinite – knows all – but confusedly. It is like walking on the seashore and hearing the great noise of the sea: I hear the particular noises of each wave, of which the whole noise is composed, but without distinguishing them. It is the same when it comes to each monad (Leibniz, *Principes de la nature de de la grâce*, § 13, GP VI, 604, trans. AG, 211 (modified)).

Everything that is within the clear zone of perception of a given monad, will be perceived as an integrated distinct macro-perception, whereas all the perceptions that fall outside that clear zone of perception – that is, the monad's perception of the rest of the world – remain unintegrated and confused perceptions, like a perceptive murmur in the back of our head that relates the rest of the world to us confusedly, a 'differential unconscious' as Deleuze himself calls it.[39]

This is what is aptly illustrated by the convergence and divergence of normal lines in relation to the position of the monads m(P) and m(Q) in relation to the curve: where the normal lines converge or meet in the *situs* of a given monad, perception of the corresponding events on the curve are

---

[39]This will prove salutary for Leibniz's philosophy in the context of Deleuze's own Nietzschean position, and explains in some measure why Deleuze occasionally is willing to label himself a 'Leibnizian': the 'Dionysian' rumbles in the depths of the differential unconsciousness of minute perceptions. On Deleuze's conception of the differential unconscious, see *Le Pli*, 114–7, and *Différence et répétition*, 214, 275–6.

integrated into clear macro-perception in that monad, for example, the events p1, p2, and p3 in relation to m(P); where the normal lines diverge from that *situs*, on the contrary, perception in the monad is differentiated into obscure minute perception, for example, the events q1, q2, and q3 in relation to m(P). We should finally note how this notion of clear zones of expression is yet another way of formulating a principle of individuation for monads. For since two monads cannot inhabit the same segment of the fold, each monad necessarily has a clear zone of expression that is different from that of any other: 'Monads all go confusedly to infinity, to the whole; but they are limited and differentiated by the degrees of their distinct perceptions' (Leibniz, *La Monadologie*, § 60, GP VI, 617, trans. AG, 221).

10. CONCLUSION

I have attempted to describe in an accessible way how Deleuze's notion of folding captures the basic construction of Leibniz's monadological metaphysics. I have done so by means of five figures, all referring to the same basic figure of a fold. I do not pretend to have given all the fold figures one might imagine for explaining Deleuze's particular take on Leibniz's monadology. There is, I suspect, more mileage in the exercise. I have, for example, not touched upon the problem of how Leibniz's conception of monadic domination in organic bodies or his difficult theory of substantial bonds may be illustrated by elaborating further the fold-figure, although I am convinced that might very well be possible. Conversely, I do not pretend either that, simply by multiplying the figures further, all aspects of Leibniz's metaphysics could be accounted for, or even that the figures I *have* provided are fully adequate to illustrate entirely the aspects they are intended to capture. Nonetheless, these figures can help, I hope, alleviate some of the difficulties in understanding Deleuze's approach to the *Monadology* from the non-Deleuzian perspective of contemporary Leibniz scholarship and give a sense of the synthetic, explanatory force that Deleuze's notion of folding has in relation to Leibniz's monadological metaphysics.[40]

BIBLIOGRAPHY

Aristotle. *The Complete Works of Aristotle*. Edited by J. Barnes. Princeton: Princeton University Press, 1984.

---

[40] I am grateful to Paul Lodge and to the anonymous reviewers of the BJHP for numerous corrections and suggestions that much helped me improve this paper.

Arthur, Richard T. W. 'Leibniz in Continuity'. *PSA: Proceedings of the Biennial Meeting of the Philosophy of Science Association*, no. 1 (1986): 107–15.
Arthur, Richard T. W. *Leibniz*. Oxford: Polity Press, 2014.
Benjamin, Walter. *The Origin of German Tragic Drama*. Translated by J. Osborne. London: Verso, 1998.
Bouquiaux, Laurence. 'Plis et enveloppements chez Leibniz'. In *Différence et identité*, edited by G. Cormann, S. Laoureux and J. Pieron, 39–56. Hildesheim: Georg Olms Verlag, 2005.
Bourbaki, Nicolas. 'L'architechture des mathématiques'. In *Les grands courants de la pensée mathématique*, edited by F. Le Lionnais, 35–47. Paris: Hermann, 1948.
Conley, Tom. 'Translator's Foreword. A Plea for Leibniz'. In *The Fold. Leibniz and the Baroque*. Translated and edited by G. Deleuze and Translated by T. Conley, ix–xx. London: Athlone Press, 1993.
Couturat, Louis. 'Sur la métaphysique de Leibniz (avec un opuscule inédit)'. *Revue de Métaphysique et de Morale* 10, no. 1 (1902): 1–25.
Deleuze, Gilles. *Logique du sens*. Paris: Minuit, 1669.
Deleuze, Gilles. *Différence et répétition*. Paris: Presses Universitaires de France, 1968.
Deleuze, Gilles. *Spinoza et le problème de l'expression*. Paris: Minuit, 1968.
Deleuze, Gilles. 'À quoi reconnaît-on le structuralisme?' In *Histoire de la philosophie VIII. Le XXe siècle*, edited by F. Châtelet, 299–335. Paris: Hachette, 1973.
Deleuze, Gilles. *Logique de la sensation*. Paris: Éditions de la Différence, 1981.
Deleuze, Gilles. *L'Image-mouvement. Cinéma 1*. Paris: Minuit, 1983.
Deleuze, Gilles. *L'Image-temps. Cinéma 2*. Paris: Minuit, 1985.
Deleuze, Gilles. *Le Pli. Leibniz et le Baroque*. Paris: Minuit, 1988.
Deleuze, Gilles. *Pourparlers*. Paris: Minuit, 1991.
Duffy, Simon. 'Deleuze, Leibniz and Projective Geometry in *The Fold*'. *Angelaki. Journal of the Theoretical Humanities* 15, no. 2 (2010): 129–47.
Fenves, Peter. 'Autonomasia: Leibniz and the Baroque'. *MLN* 105, no. 2 (1990): 432–52.
Fichant, Michel. 'Actiones sunt suppositorum. L'ontologie leibnizienne de l'action'. *Philosophie* 53 (1997): 135–48.
Fichant, Michel. 'L'Invention métaphysique'. In *Discours de métaphysique, Monadologie*, edited by G. W. Leibniz and M. Fichant, 7–140. Paris: Gallimard, 2004.
Garber, Daniel. *Leibniz. Body, Substance, Monad*. Oxford: Oxford University Press, 2009.
Knecht, Herbert. *La Logique chez Leibniz: essai sur le rationalisme baroque*. Lausanne: l'Âge d'homme, 1981.
Kulstad, Mark: 'Leibniz's Conception of Expression'. *Studia Leibnitiana*, no. 9 (1977): 56–76.
Lærke, Mogens. 'Four Things Deleuze Learned from Leibniz'. In *Deleuze and the Fold: A Critical Reader*, edited by N. McDonnell and S. Van Tuinen, 25–45. Hampshire: Palgrave MacMillan, 2009.

Lærke, Mogens. 'Compossibility, Compatibility, Congruity'. In *Leibniz: Compossibility and Possible Worlds*, edited by Y. Chiek and G. Brown. Dordrecht: Springer, Forthcoming.

Leibniz, Gottfried Wilhelm. *Leibnizens mathematische Schriften*. Edited by C. I. Gerhardt. Berlin: A. Asher, 1849–60.

Leibniz, Gottfried Wilhelm. *Sämtliche Schriften und Briefe*. Berlin: Akademie Verlag, 1923.

Leibniz, Gottfried Wilhelm. *Die philosophischen Schriften von Gottfried Wilhelm Leibniz*. Edited by C. I. Gerhardt. Hildesheim: Georg Olms Verlag, 1965.

Leibniz, Gottfried Wilhelm. *Philosophical Essays*. Translated and edited by R. Ariew and G. Garber. Indianapolis: Hackett, 1989.

Leibniz, Gottfried Wilhelm. *Philosophical Papers and Letters*. Translated and edited by L. E. Loemker. Dordrecht: Kluwer, 1989.

Leibniz, Gottfried Wilhelm. *The Labyrinth of the Continuum. Writings on the Continuum Problem, 1672– 1686*. Translated and edited by R. W. T. Arthur. New Haven: Yale University Press, 2001.

Leibniz, Gottfried Wilhelm. *The Leibniz-Des Bosses Correspondence*. Translated and edited by B. Look and D. Rutherford. New Haven: Yale University Press, 2007.

Leibniz, Gottfried Wilhelm. *Protogaea*. Translated and edited by C. Cohen and A. Wakefield. Chicago: Chicago University Press, 2008.

Levey, Samuel. 'The Interval of Motion in Leibniz's Pacidius Philalethi'. *Noûs* 37, no. 3 (2003): 371–416.

McDonnell, Niamh, and Sjoerd Van Tuinen, eds. *Deleuze and The Fold. A Critical Reader*. Basingstoke: Palgrave MacMillan, 2010.

Mercer, Christia. *Leibniz's Metaphysics*. Cambridge: Cambridge University Press, 2001.

Peden, Knox. *Spinoza Contra Phenomenology. French Rationalism from Cavaillès to Deleuze*. Stanford: Stanford University Press, 2014.

Robinson, Keith. 'Events of Difference: The Fold in Between Deleuze's Reading of Leibniz'. *Epoché* 8, no. 1 (2003): 141–64.

Smith, Daniel W. 'Deleuze on Leibniz: Difference, Continuity, and the Calculus'. In *Current Continental Theory and Modern Philosophy*, edited by S. H. Daniel, 127–47. Evanston: Northwestern University Press, 2006.

Swoyer, Chris. 'Leibnizian Expression'. *Journal of the History of Philosophy* 33, no. 1 (1995): 65–99.

ARTICLE

LEIBNIZ'S MONADOLOGICAL POSITIVE AESTHETICS

Pauline Phemister and Lloyd Strickland

One of the most intriguing – and arguably counter-intuitive – doctrines defended by environmental philosophers is that of positive aesthetics, the thesis that all of nature is beautiful. The doctrine has attained philosophical respectability only comparatively recently, thanks in no small part to the work of Allen Carlson, one of its foremost defenders. In this paper, we argue that the doctrine can be found much earlier in the work of Gottfried Wilhelm Leibniz who devised and defended a version of positive aesthetics (*avant la lettre*) in the early modern period, grounded in a conception of the world as a world of monads, each of which individually fulfils the rationalist aesthetic criteria of multiplicity-in-unity and that taken together ensure that the world as a whole is a harmoniously ordered system of multiple and diverse individuals, whose intelligible order and variety is made known to us through natural scientific endeavour. In showing this, we advance two further theses: first, that Leibniz's version of positive aesthetics displays more philosophical virtue than Carlson's, for whereas Carlson's doctrine is vague and admits of exceptions, Leibniz's is clear and all-encompassing. And secondly, that Leibniz's version of positive aesthetics has the resources to overcome a difficulty inherent in the exclusively science-based justification that Carlson offers.

INTRODUCTION

One of the most intriguing – and arguably counter-intuitive – doctrines defended by environmental philosophers is that of positive aesthetics, the thesis that all of nature is beautiful. A perusal of the literature on the topic would lead one to suppose that while the belief that all nature is beautiful has been held by a number of artists and naturalists over the last two centuries,[1] it has attained philosophical respectability only comparatively recently, thanks in no small part to the work of Allen Carlson, one of its

---
[1] See Carlson (*Aesthetics and the Environment*, 73–4) for examples.

foremost defenders. In this paper, we argue that the doctrine can be found much earlier in the work of Gottfried Wilhelm Leibniz, who devised and defended a version of positive aesthetics (*avant la lettre*) in the early modern period, grounded in a conception of the world as a world of monads, each of which individually fulfils the rationalist aesthetic criteria of multiplicity-in-unity and that taken together ensure that the world as a whole is a harmoniously ordered system of multiple and diverse individuals, whose intelligible order and variety is made known to us through natural scientific endeavour. In showing this, we advance two further theses: first, that Leibniz's version of positive aesthetics displays more philosophical virtue than Carlson's, for whereas Carlson's doctrine is vague and admits of exceptions, Leibniz's is clear and all-encompassing. And second, that Leibniz's version of positive aesthetics has the resources to overcome the difficulty inherent in the science-based justification that Carlson offers. To show this, we shall first outline Carlson's doctrine of positive aesthetics, and then turn our attention to Leibniz.

CARLSON'S POSITIVE AESTHETICS

In *Aesthetics and the Environment*, Allen Carlson defines positive aesthetics as the 'initially implausible' view that 'all the natural world is beautiful' and that 'the natural environment, insofar as it is untouched by humans, has mainly positive aesthetic qualities' (Carlson, *Aesthetics and the Environment*, 72).[2] We might wonder – as indeed have others, such as Malcolm Budd – whether Carlson is here thinking of nature as whole, or the biosphere, ecosystems, species, individual (natural) things, or individual (natural) events.[3] While Carlson does not specify, the way in which he seeks to

---

[2]Originally published as Carlson 'Nature and Positive Aesthetics'. As we shall see over the course of this paper, Leibniz's belief that the whole of nature is beautiful differs in various ways from the position that Carlson adopts. But one difference is worth noting now, which is this: Leibniz makes a distinction between the living machines of nature created by God and the artificial machines constructed by humans. The former have infinitely many enfolded parts; the latter have only a finite number of parts (*New System*: GP IV 482; L 456). The organic bodies of human beings are living machines and as such are just as much a part of nature as any other living body. Therefore to consider nature as beautiful only 'insofar as it is untouched by humans' is to separate the natural from the human in a way that Leibniz would not endorse.

[3]See Budd (*Aesthetic Appreciation of Nature*, 97). Needless to say, there is a rich literature on positive aesthetics, and in it one can find various other objections to the doctrine, or at any rate to versions thereof. It is beyond the scope of this paper to offer a detailed examination of this literature, and the objections found therein, and we thus leave it to the reader to assess how well Leibniz's version of positive aesthetics fares against objections found (for example) in Fisher 'Aesthetics', 271–3 and Brady 'Ugliness and Nature'. It should be noted, however, that many objections to positive aesthetics are aimed at one specific version of the doctrine. For example, the objection that it is highly implausible to suppose that all parts and elements of nature are equally beautiful (articulated, e.g. in Budd, *Aesthetic Appreciation of Nature*, 127

justify positive aesthetics affords us a clue. He argues that in our appreciation of nature, we should adopt what he calls the 'natural environment model', according to which 'we must appreciate nature ... in light of knowledge provided by the natural sciences, especially the environmental sciences such as geology, biology and ecology' (Carlson, *Aesthetics and the Environment*, 6). More specifically, we should appreciate natural things under their correct categories, which are the categories that are discovered by naturalists. So for example, if we are appreciating a whale, we should do so under the category of 'mammal' rather than 'fish' or 'bird' or 'marsupial' etc. (Carlson, *Aesthetics and the Environment*, 63ff). This enables us to appreciate it as what it actually is, which in turn enables us to have a richer, deeper appreciation of that thing. Unfortunately Carlson's insistence on our appreciating natural things under their correct categories does not enable us to determine precisely what he means when he talks of all nature having only positive aesthetic qualities, though it does allow us to make an educated guess. Certainly Carlson's choice of illustration for the doctrine (individual whales) suggests he thinks it applies to individual natural things (e.g. individual whales). However, we should here note Carlson's concession that some natural things are *not* aesthetically good. When faced with the objection that 'grossly malformed living things will remain grotesque no matter how comprehensible science renders their malformation' (Budd, *Aesthetic Appreciation of Nature*, 102, cf. 126), Carlson conceded that grossly malformed living things are exceptions to his view on positive aesthetics (Carlson, 'Hargrove, Positive Aesthetics, and Indifferent Creativity', 234n36). In another text, he admitted that 'damaged, diseased, and malformed living things' are all exceptions (Parsons and Carlson, *Functional Beauty*, 136).[4] Clearly, then, Carlson sees positive aesthetics applying to individual natural things, but not to all of them, and not in all of their states. We might reasonably suppose that he would take it to apply to individual natural events as well, but it is less clear that he would also apply it to nature as whole, the biosphere, ecosystems, or species, given that these do not, according to conventional wisdom, fall under the senses because of their vastness or abstractness. But there is an element of conjecture here, because ultimately Carlson leaves the parameters of his version of positive aesthetics rather vague.

In any case, even if Carlson is unclear about what exactly falls under his version of positive aesthetics, he is clear about how it is to be grounded, insisting that scientific study of the natural world is the most effective

---

and Fisher 'Aesthetics', 272) has force only against versions of positive aesthetics that make such a claim.

[4]In this book, the authors advance an aesthetics of nature that emphasizes the functional values that individuals possess in relation to each other and their environments. Although Leibniz's metaphysical views are highly relevant to this approach, it would take us too far from the topic of this paper to explore them here.

route to the discovery of the positive aesthetic qualities that nature possesses in itself:

> When nature is aesthetically appreciated in virtue of the natural and environmental sciences, positive aesthetic appreciation is singularly appropriate, for, on the one hand, pristine nature – nature in its natural state – is an aesthetic ideal and, on the other, as science increasingly finds, or at least appears to find, unity, order, and harmony in nature, nature itself, when appreciated in light of such knowledge, appears more fully beautiful.
> (Carlson, *Aesthetics and the Environment*, 12)

Insofar as scientific investigation of nature uncovers the aesthetic qualities that the natural world possesses in itself, it also serves as justification of a positive aesthetics of nature. The aim of the sciences is the intelligible explanation of phenomena. And in this endeavour the natural sciences seek to explain natural phenomena in terms of 'order, harmony, balance, tension, resolution, and so forth' (Carlson, *Aesthetics and the Environment*, 93). These, as Carlson explains, are precisely the qualities that we find aesthetically pleasing:

> these qualities that make the world seem comprehensible to us are also those that we find aesthetically good. Thus, when we experience them in the natural world or experience the natural world in terms of them, we find it aesthetically good. This is not surprising, for qualities such as order, regularity, harmony, balance, tension, and resolution are the kinds of qualities that we find aesthetically good in art.
> (Carlson, *Aesthetics and the Environment*, 93)

Carlson does not speculate as to why it is that the criteria we apply in the sciences are the same criteria against which we judge the aesthetic qualities of works of art (ibid.). However, what is important in respect of a positive aesthetics of nature is that the scientific criteria lead to the discovery of truths about the natural world. It is not just that scientific studies are guided by the desire to make the natural world intelligible to us in terms of order, harmony, balance, and the like, but also that in doing so, science discovers that the natural world is a world in which these aesthetically appealing qualities really are present. The natural sciences, he contends, do not impose these qualities on the environment. Rather, scientific investigation brings to light what is already there. In this way, the sciences reveal that nature itself is aesthetically good. In Carlson's own words,

> these categories not only make the natural world appear aesthetically good, but in virtue of being correct determine that it is aesthetically good.
> (Carlson, *Aesthetics and the Environment*, 94)

The more that the sciences succeed in making intelligible the natural world as a world that is ordered, regular, harmonious and balanced, the more that world is found to be not only intelligible, but also aesthetically pleasing: 'the development of science and its continual self-revision ... constitutes a movement that puts the natural world in an increasingly favorable aesthetic light' (Carlson, *Aesthetics and the Environment*, 94–5).

Furthermore, Carlson argues that the sciences provide the only viable justification for positive aesthetics. Of particular relevance to our purposes here is Carlson's rejection of theological justification as unsatisfactory. As Carlson sees it, a theological justification for positive aesthetics would be founded upon the belief that, having been designed and created by a perfect God, the world too would be perfect and, as such, all its aesthetic qualities would be positive qualities. While Carlson acknowledges that such an appeal to divine perfection would appear to justify the doctrine of positive aesthetics, he goes on to identify three 'puzzles' connected with it (Carlson, *Aesthetics and the Environment*, 82). First, insofar as the justification is unavailable to atheists, it suggests, although it does not necessitate, the somewhat counter-intuitive notion that atheists and theists appreciate nature in radically different ways. Second, traditional appeals to the perfection of the world give rise to the 'problem of evil', which theists usually tackle not by denying the existence of evil, but by constructing a theodicy that seeks to explain why God would permit evil to exist in the world. Yet in the analogous 'problem of ugliness', which would aim to reconcile God's existence with that of ugliness in the natural world, theists seeking to justify positive aesthetics through theology would be offering a decidedly non-analogous kind of solution, in that they would be simply denying the existence of ugliness, rather than attempting to create an aesthetic theodicy. Carlson considers this incongruous, arguing that the tradition of developing theodicies to tackle the problem of evil suggests that the theist should develop an aesthetic theodicy for the problem of ugliness rather than just denying the existence of ugliness.[5] Third, and finally, Carlson points to the paucity of historical evidence of Christian thought promoting the appreciation of nature. On the contrary, he insists, Christianity has 'traditionally viewed wild nature as something to be confronted, dominated, and domesticated by human beings for their purposes' (Carlson, *Aesthetics and the Environment*, 83) and that, at least in the West, 'Christianity and the aesthetic appreciation of nature have been opposing forces to an extent that the latter could grow only as the former went into decline' (Carlson, *Aesthetics and the Environment*, 84).

In our view, Carlson's concerns about the theological justification of positive aesthetic are not warranted. For instance, his first point – that the theological justification implies that theists and atheists appreciate nature differently – is not necessarily problematic. The difference could simply be acknowledged and accepted or absorbed within a wider conception of

---

[5]Although he does not put it quite like this, Carlson's point seems to be that analogous problems should admit of analogous solutions.

the role of theistic and atheistic appreciations of nature.[6] Regarding the second objection, Carlson's suggestion that because theists have engaged in theodicy to resolve the problem of evil they should likewise engage in aesthetic theodicy to resolve the problem of ugliness (rather than simply denying ugliness), is clearly under-motivated. There is no compelling reason why the two problems cannot be resolved in different ways. But if one insists that they have to be, one need only look at the work of Leibniz to see how both problems can be resolved in analogous ways. For Leibniz acknowledged evil in the parts while insisting that such evil is a necessary feature of a perfect whole and disappears in relation to it;[7] similarly he proposes (as we shall see) that negative aesthetic qualities in nature are either merely apparent or real only when the individual parts of nature are considered in isolation from the whole. Leibniz thus offers a traditional theodicy, and a cognate aesthetic theodicy. Finally, to Carlson's third objection, the historical fact of Christianity's appalling track record is not in dispute: for centuries, the basic thrust of Christianity has been disadvantageous to the non-human world. However, historical record is not particularly relevant to the question whether it is possible to justify positive aesthetics of nature on theological grounds. The fact that Christians have historically failed to acknowledge the goodness and beauty of the natural world is not a declaration of the inevitability of Christian antipathy towards nature. The future need not always resemble the past. In any case, it is far from clear that the failure is as great as Carlson makes it out to be: the historical record affords us numerous examples of Christian thinkers who have readily made positive aesthetic judgements about nature, such as Thomas Burnet, John Ray, and George Berkeley.[8] And today, positive, nature-affirming interpretations of Scripture are available as part of a contemporary theological drive towards a re-evaluation of the relation of God and world. Irrespective of how we view the historical record, in recent times, writings such as those of prominent theologian Jürgen Moltmann stand testimony to the power of Christian thought to support human appreciation and concern for Creation (Moltmann, *God in Creation*).

All the same, even if the objections to the theological defence of positive aesthetics obtain and theological appeals to God's perfection and creativity are found insufficient to ground a positive aesthetics of nature, this does not mean that theological considerations should be rejected *tout court*, for, as will be argued below, they do have important justificatory value.

We begin, however, with a defence of a positive aesthetics interpretation of Leibniz's account of the beauty of the natural world, before going on to

---

[6]The conclusions to be drawn later in this paper would support this second suggestion.
[7]See, for example, Leibniz's *Remarks on the three volumes entitled Characteristics of Men, Manners, Opinions, Times, ... 1711*: GP III 429; L 633. See also *On the Ultimate Origination of Things*: GP VII 306–8; AG 153–4.
[8]See Burnet, *Sacred Theory of the Earth*, 109; Ray, *Wisdom of God Manifested in the Works of Creation*, 150ff; and Berkeley, *Philosophical Writings*, 192. And Leibniz is of course another who makes positive aesthetic judgements about nature, as we shall show.

explore the role of the natural sciences in bringing the objective beauty of nature into view. We will find that the natural sciences, while a necessary staging-post, are incapable of providing a full justification of positive aesthetics. We will also discover that, for Leibniz, not only do the sciences themselves require theological grounding, but also that a theological justification of natural beauty uncovers a world founded upon monads, the extent of whose beauty is even greater than that which the natural sciences can reveal. In conclusion, it will be suggested that the most solid defence of positive aesthetics lies in the combination of a scientific justification underpinned and extended by a rationalist theology. Only when the sciences and theology work together does positive aesthetics come to rest on a sure foundation.

THE POSITIVE AESTHETICS OF LEIBNIZ

Following Ernst Cassirer (Cassirer, *Philosophy of the Enlightenment*, 34), recent accounts of eighteenth-century rationalist aesthetics have acknowledged the undoubted immense debt to Leibniz owed by his immediate successors, Christian Wolff and Alexander Baumgarten. For instance, Frederick Beiser (Beiser, *Diotima's Children*) and Paul Guyer (Guyer, *A History of Modern Aesthetics*, 49–52) have drawn our attention to the influence of Leibniz's thought on the eighteenth-century rational aesthetics that stem from Wolff and Baumgarten. The genealogy is instructive and persuasive. As Beiser's penetrating and lucid account shows, Leibniz's Principle of Sufficient Reason provided the basis for an aesthetics that maintains that the predication of beauty must be grounded upon intelligible principles of harmony, order, and perfection, understood in terms of 'unity-in-variety' (Beiser, *Diotima's Children*, 4–8). Nevertheless, there are dangers and pitfalls that must be avoided when we read Leibniz backwards through the prism of the tradition of rationalist aesthetics that his thought inspired. The aesthetics of Wolff and Baumgarten combine both subjective and objective elements. For Wolff, beauty is subjective insofar as it exists as a feeling of pleasure: if there is no feeling of pleasure, there is no beauty. But insofar as this feeling of pleasure is a response to an actual perfection in the object itself, beauty is also objective. The subjective and objective elements are combined, Beiser holds, in Wolff's definition of beauty as the 'observability of perfection': the definition 'neatly joins both these elements together, for it means that beauty is neither perfection nor pleasure alone but both: the pleasure from observing perfection' (Beiser, *Diotima's Children*, 63). On Beiser's reading, Baumgarten adopts the same stance: beauty is objective insofar as it is grounded in the actual perfection of the object and it is subjective insofar as beauty must be sensed and cognized.[9]

---

[9]For Baumgarten, aesthetics is the 'science of sensitive *cognition*' (*Aesthetics*, §1). See Beiser, *Diotima's Children*, 119.

> Following Wolff, Baumgarten's central thesis is that beauty consists in the intuition of perfection. ... Such a thesis attempts to explain both the subjective and objective aspects of beauty. In making perfection essential to beauty, it makes beauty partially objective. If there were no variety-in-unity in the object, there would be no beauty. But in making intuition also crucial to beauty, it also makes beauty subjective. If there were no sensible perception of perfection, there also would be no beauty.
>
> (Beiser, *Diotima's Children*, 145)

Looking back to Leibniz from the perspective offered by Wolff and Baumgarten, it is natural to read Leibniz in the same way. Accordingly, Beiser states that for Leibniz, 'beauty is both a subjective and objective quality' (Beiser, *Diotima's Children*, 36). As evidence, he cites Leibniz's well-known definition of beauty as 'that, the contemplation of which is pleasant' (*Elements of Natural Law*: A VI i 464: L 137). On Beiser's reading, 'contemplation' introduces a subjective element into Leibniz's account of beauty that is additional to the objective element already present insofar as Leibniz regards pleasure as a feeling that arises when perfections that exist objectively in the things are perceived (Beiser, *Diotima's Children*, 35).

However, when we read Leibniz on his own terms, freed from the legacy of Wolff and Baumgarten's interpretations, the case for reading Leibniz's definition as indicative of the subjectivity of beauty is significantly weakened. Reading Leibniz not through his enlightenment reception, but rather through the lens of his classical heritage brings Leibniz's opinion that beauty is an objective quality into sharper focus. The general consensus prior to the eighteenth century was that the objective beauty of things was found in the unified order and proportion of their parts,[10] and therefore does not depend upon its also being subjectively perceived. Under this light, Leibniz's claim that it is pleasing to contemplate beautiful things no more suggests that things are beautiful only when we actually contemplate them than it suggests that the pleasure we get from contemplating or discovering scientific truths suggests that they are true only when we are actually contemplating them. When introducing his definition of beauty as 'that, the contemplation of which is pleasant', he prefaced his remark with the observation that 'We seek beautiful things because they are pleasant' (*Elements of Natural Law*: A VI i 464: L 137). This strongly suggests that beauty exists independently of our perception or contemplation of it and that the contemplation of beautiful things is pleasing to us in consequence of their beauty, not a requisite for their beauty. They would be beautiful even were there no one to contemplate them. The same is true of God's perception. God

---

[10] Augustine, for instance, endorsed the by then popular Aristotelian view that conceives beauty in terms of the orderly arrangement of parts. See Aristotle, *Poetics*, 1450a36-37: Barnes, *Complete Works of Aristotle*, II 2322; Augustine, *De ordine* II.15.42: Borruso, *St. Augustine: On Order (De Ordine)*, 105–7.

knows the beauty of things, but they are beautiful in themselves, not beautiful because God knows them.[11]

Other of Leibniz's remarks on the beauty of things confirm what is here implied, namely that beauty is wholly objective, and not reliant in any way on subjective perception or contemplation.[12] For example, in a piece on true piety and the love of God, Leibniz writes that 'we consider a painting excellent not because of some usefulness to us, but because of *its own beauty*' (The elements of true piety, or, on the love of God over everything: A VI iv 1357: SLT 189, our emphasis), while shortly after in the same piece he repeats his observation that '*The beautiful* is that, the contemplation of which is pleasant' (A VI iv 1358: SLT 190). Were this remark to be read as laying claim to the subjectivity of beauty, it would undermine Leibniz's earlier comment in the same piece on the objective beauty of the painting. Leibniz also holds that God Himself is beautiful – 'nothing is happier than God and also nothing can be understood as being more beautiful or more worthy of happiness' (Preface to the *Diplomatic Code of People's Rights*: GP III 387: SLT 150). God's beauty depends neither on our, nor on God's, contemplation of it. Finally, we may note that it is not only individual created things and God that are objectively beautiful. Leibniz asserts that the world as a whole has an objective beauty that, far from being even in part subjectively dependent upon our contemplation or pleasure, already exists, ready for us to discover provided we pay sufficient attention informed by the truths of mathematics:

> we must acknowledge that it is important that one have some general insights on mathematics, not as craftsmen have for the accuracy of their works, but because of the openings that one finds in it for elevating the mind to thoughts that are beautiful and sound in equal measure. For without that the items of human knowledge are only vague and superficial. This is clearly seen with regard to the system of the visible universe, about which the previous century and ours have made wonderful discoveries, and what the ancients knew of it was mere juvenilia compared to what is known about it now. This system or structure of the visible world is of an admirable beauty which gives true ideas of the grandeur and harmony of the universe ...
> (Leibniz to Sophie, 23 October/2 November 1691: A I vii 49–50: LTS 91–2)

Passages such as these portray Leibniz as an early exponent of a positive aesthetic of nature. First, he assumes that beautiful things are objectively beautiful. Their beauty does not depend upon being felt or contemplated. Beauty is a feature of things themselves, discoverable through close attention to the empirical detail, grounded in the cognitive truths of the mathematical sciences. Second, as we see from the passage just cited from Leibniz's

---

[11]See, for example, *Rationale Fidei Catholicae*, A VI iv, 2320; LGR 76–7.
[12]Obviously, of course, subjective perception or contemplation is still necessary if this objective beauty is to be perceived and appreciated.

correspondence with Sophie, Leibniz also views nature or the universe in a wholly positive light. Other passages are even more explicit about this; for example, Leibniz informs us that 'God created everything in accordance with the greatest harmony or beauty possible' (*Aphorisms concerning happiness, wisdom, charity and justice*: A VI iv 2799; LGR 138). Thus, in anticipation of rational aesthetics in the eighteenth century and Carlson's adoption of similar criteria for aesthetic goodness in our own, Leibniz held that the true perfection and beauty of the universe resides in its being a harmoniously ordered and infinitely varied plurality of individual living substances. Any ugly disorder we believe to occur in nature is more apparent than real. When we see such negative qualities in the context of the whole, the disorder vanishes:

> the apparent disorders are only like certain chords in music which sound bad when one hears them by themselves, but which a skillful composer leaves in his work because by combining them with other chords they increase one's enjoyment, and render the whole harmony more beautiful.
> (Leibniz to Sophie Charlotte, 9/19 May 1697: GP VII 545: LTS 160)

Elsewhere, Leibniz explains that 'all the imperfections we think we find in the world only originate from our ignorance' and that we lack 'the right point of view to judge of the beauty of things' (to André Morell, 29 September 1698: A I xvi 162: SLT 197). We have relatively distinct perceptions of only a tiny fragment of this spatially and temporally infinite plenum of a universe. For the most part, our perceptions are confused: we do not clearly perceive the parts of an object, or of the universe, and so fail to appreciate fully and intellectually the perfect harmonious order of the whole. It is hardly surprising, therefore, that we fail to appreciate its beauty in all its finest glory (Leibniz to Sophie Charlotte, 9/19 May 1697: GP VII 545: LTS 160). Nevertheless, even when our perceptions are confused, we may still have a sense that 'There is something, I know not what, that pleases me in the matter,' and this sense testifies to the fact that at the level of 'our feelings [*Gemüth*]' we have registered or perceived the presence of intrinsic, objective perfection and beauty (*On Wisdom*: GP VII 86: L 425–6).

LEIBNIZ AND THE SCIENCES

One might with reason expect Leibniz to justify his positive aesthetics theologically by simply appealing directly to God's goodness and perfection. After all, Leibniz does maintain that God freely chose to create this world rather than any other world because this world is the best of all possible worlds. Obviously, an omnipotent, omniscient, and benevolent God, wisely guided by the principle of the best, will choose to create that world that is the most harmonious, most ordered, good, and beautiful, in short, the world that is the most perfect.[13]

Surprisingly, however, Leibniz does not, at least in the first instance, justify the positive beauty and perfection of the world by appealing to God's perfection. On the contrary, we find him arguing in the opposite direction: the beauty and perfection that we find in the world leads us to knowledge of God. Leibniz declares that there are in fact two ways that lead to knowledge of God's beauty and perfections 'through his emanations':

> namely in the knowledge of eternal truths, explaining the reasons in themselves, and in the knowledge of the harmony of the universe, by applying reasons to facts. That is to say, we must know the wonders of reason and the wonders of nature.
> (Happiness: Gr 580–1; SLT 168)

The 'wonders of reason and of eternal truths' concern the truths of arithmetic, geometry, justice, and morals that the mind 'discovers in itself in the sciences of reasoning'. However, the 'wonders of corporeal nature', which include 'the system of the universe, the structure of the bodies of animals, the causes of the rainbow, of the magnet, of tidal ebb and flow, and a thousand other similar things' (Happiness: Gr 581; SLT 168), had been discovered only through natural scientific application of 'reasons to facts'. It was this methodology that had led to Nicolas Copernicus's revolutionizing of our understanding of the heavens, Johannes Kepler's discovery of the elliptical paths of the planets and realization of the role played by the moon in the turning of the tides, the painstaking uncovering of minute worlds by microscopists such as Antonie van Leeuwenhoek, William Harvey's discovery of the circulation of the blood, Descartes's and Newton's studies of the rainbow, and William Gilbert's theory of magnetism.

Although he does not coin the phrase, in Leibniz's acknowledgment of the value of the applied natural sciences in bringing to light the 'wonders of nature', we see him moving towards a justification of positive aesthetics that bears a remarkable affinity to Carlson's own justification by the natural sciences. From mathematical physics and astronomy to chemistry, anatomy, and physiology, the natural sciences were transforming early modern understanding of the natural world and humans' place within it. But more than this, they were discovering order, variety, and harmony throughout all parts of the universe. And the more they discovered, the more the beauty of the universe came to light. In the words of Leibniz:

> It is only in our time that we are beginning to recognize the secret of both the little and the great world, by the discovery, on the one hand, of the circulation

---

[13] See *Theodicy* §416: G VI 364, where Leibniz envisions possible worlds being ranked in terms of beauty, and God choosing the most beautiful one, namely ours, the best of all possible worlds.

of the blood in ourselves, and on the other hand (by means of telescopes) of the true movements of the heavenly bodies. If human beings continue to make progress as they have within the past hundred years, many things of wonderful beauty will be displayed by nature ...
(Thoughts on Van Helmont's doctrines, first half of October (?) 1696: A I xiii, 51; LTS 139)

Leibniz's stance is clearly in the spirit of Carlson's natural scientific defence of positive aesthetics. There are differences in their respective conceptions of the natural sciences: Leibniz's mathematically informed approach stands in contrast to the natural historical approach favoured by Carlson in which classification according to natural categories dominates.[14] Nevertheless, whatever the actual methodology employed, both Leibniz and Carlson are convinced that natural scientific investigation is the means by which objective natural beauty, as intelligible order and harmony, is discovered and appreciated.

Leibniz himself, however, goes beyond this to claim that those 'many things of wonderful beauty' that we find in nature, 'give us yet more cause to esteem their creator, and to take pleasure in his acts' (Thoughts on Van Helmont's doctrines, first half of October (?) 1696: A I xiii, 51; LTS 139). The discovery of beauty in nature leads us to acknowledge not only the positive beauty of the natural world, but also the wisdom, perfection, and beauty of the creator and of the individual perceivers of this beauty. As he remarked to Sophie Charlotte, the more we uncover the order and harmony in the natural world, the more we are convinced that

> the universe is governed by a sovereign intelligence, in an order so perfect that, if one understood it in detail, one would not only believe but would even see that nothing better could be wished for.[15]
> (Leibniz to Sophie Charlotte, 9/19 May 1697: GP VII 545; LTS 160)

---

[14]We thank an anonymous referee for this point. We thank another anonymous referee for pointing out that there is a degree of subjectivity inherent in the human categorization of nature that creates a certain tension with the idea of objective beauty grounded in these categories. The idea is that the categories we employ in our understanding of nature are to some extent nominal or arbitrary. Carlson himself, however, considers the categories of nature as objectively true natural kinds, discovered rather than created by natural sciences (Carlson, *Aesthetics and the Environment*, 90). The issue does not arise for Leibniz, for he locates the beauty of nature not in categories or species, but in its unique individual constituents that together form a similarly unique and beautiful whole.

[15]Indeed, Leibniz even saw in the beauty of the world an argument – albeit not a demonstrative one – for the existence of God, for he writes that 'from the beauty of things alone it is indeed very probable that the world was constructed by a most wise architect', that is, God (On freedom, fate and God's grace: A VI iv 1604; LGR 262). In one early writing, he even suggested that the world's beauty afforded an 'infinite probability, or moral certainty' that a mind – God's – was behind it (Sketch of *Catholic Demonstrations*: A VI i 494; LGR 22).

The natural sciences tell us *how* things work, but not *why* they work as they do. Laws and regularities are discovered, but the sciences do not and cannot explain why these ones obtain rather than others. The sciences reveal *that* the world is intelligible, but do not explain *why* it is intelligible. Nor indeed, can they offer any explanation as to *why* a world exists at all. In aesthetics, as the eighteenth-century rationalist aestheticians had agreed, Leibniz's Principle of Sufficient Reason, rigorously applied, points to the need not only to declare that we find certain things beautiful, but also to justify the attribution by giving the reasons upon which the judgement is founded. In the natural sciences, the Principle of Sufficient Reason grounds the belief that all phenomena are explicable in terms of efficient causation, but it also shows up the limitations of such explanations and highlights the need to postulate an ultimate reason that lies beyond the range of the empirical sciences. Such an ultimate reason appeals to the considerations of goodness and perfection that entered into God's decision to create the best possible world. In this, it prioritizes final causation over efficient causation: the laws of efficient causation that govern the collisions among bodies must, in the end, be explained by final causation's reference to the divine will to create that possible world that God understood to be the best (*Principles of Nature and Grace*, §§ 8–11: GP VI 602-3; AG 210–11).[16]

> We must not distrust the pleasures that arise from intelligence or reasons, when we penetrate the reason of the reason of perfections, that is to say, when we see them follow from their source, which is the absolutely perfect being.
>
> The perfect being is called God. He is the final reason of things, and the cause of causes. Being the sovereign wisdom and sovereign power, he has always chosen the best and always acts in an orderly way.
>
> (Happiness: Gr 580; SLT 168)

The natural sciences provide us in the first instance with knowledge of the beauty of nature, but in doing so they also expose their own limitations. God is needed as the ultimate reason why the world exists and the reason why the best possible world exists. God chose to create this world rather than any other *because* he knew that this world fulfilled the criteria of goodness, harmony, order, beauty, and perfection, these being attributes that this world possesses objectively, that is, in itself. We thus see that the Leibnizian theological justification of positive aesthetics works together with a scientific justification. First, examples of the beauty and perfection of the world are discovered by the natural sciences, but these in turn need to be explained by reference to decision of a perfect and rational God to create the best

---

[16] '[T]he laws of motion cannot be explained through purely geometric principles or by imagination alone ... they originate in the wisdom of their Author or in the principle of greatest perfection, which has led to their choice' (*Tentamen Anagogicum*: GP VII 271-2; L 478).

possible world, which in turn justifies the positive aesthetic claim about the whole of nature, including those parts that have not yet been uncovered through scientific investigation, and consequently of whose beauty we do not at present 'have the right point of view to judge' (to André Morell, 29 September 1698: A I xvi 162; SLT 197).

> It is a bit like in astronomy, where the motion of the planets appears to be a pure confusion when one looks at it from the Earth, but if we were in the sun we would find before our very eyes this beautiful arrangement of the system which Copernicus has discovered by dint of reasoning. As the smallest bodies are, so to speak, small worlds full of marvellous creatures, we should not imagine that there are barren parts, absolutely speaking, even though they seem barren to us.
>
> (ibid.)

Indeed, as Leibniz here hints with his reference to the absence of 'barren parts' of nature, the perfection and beauty of God's creation extends far beyond that which can ever be made known to us through the empirical sciences alone. The sciences have made incredible advances in uncovering the wealth of variety and mechanical ordering among bodies. As such, however, they are concerned only with the physical aspects of living beings, including those 'marvellous creatures' in the seemingly 'barren parts' of nature to which Leibniz referred Morell. The physical sciences struggle to describe and explain the psychical aspects of these creatures: the inner, subjective experiences had by the monadic minds, souls, or entelechies that Leibniz believed dominate and unify their organic bodies.[17] In addition to the beautiful order and variety found in bodies, there is a beautiful order and variety to be found in minds, souls and entelechies whose perceptions *'cannot be explained mechanically'* (*Monadology* §17: GP VI 609: M 17), but which instead must be explained in terms of the same kinds of reasons or final causes that inclined – but did not necessitate – God to create this world instead of any of the other possible worlds, namely the will, desire, or appetite towards what is (or what appears to be) the best. (*On the Ultimate Origination of Things*: GP VII 302: AG 150. See also *New Essays*: A VI vi 178–9; RB 178–9).

THE BEST POSSIBLE WORLD

Leibniz's vision of the best possible world describes a harmonious plenum of living beings, of perceiving souls or entelechies and their organic bodies, which bodies are themselves composed of other monads with organic

---

[17] According to Leibniz, bodies are either corporeal substances, that is, living bodies endowed with dominant minds, souls, or entelechies or aggregates of corporeal substances (to Bierling, 12 August 1711: GP VII 501).

bodies, ad infinitum. Underpinning the phenomenal world of physical nature is a metaphysical world of monadic unities. The opening sections of Leibniz's *Monadology* demonstrate the logical necessity of the indivisible, unified, and unifying monads as the foundational requisites of the physical world of divided aggregate bodies:

> *The monad*, about which we shall speak here, is nothing other than a simple substance which enters into compounds, – 'simple' meaning 'without parts'. (*Theodicy*, preliminary discourse §10).
>
> And there must be simple substances, because there are compounds; for the compound is nothing but an accumulation or *aggregate* of simples.
>
> Now where there are no parts, neither extension, nor shape, nor divisibility is possible. And these monads are the true atoms of nature and, in a word, the elements of things. (*Monadology* §§ 1–3: GP VI 607; M 14)

No created monad is ever separated from its organic body. Each is dominant over a constantly changing aggregate of substances (and aggregate of subordinate monads with their own organic bodies), forming with them an animal-like, living unity, a corporeal substance (Phemister, *Leibniz and the Natural World*, Chapters 1–3). Perceiving as a unified whole the effects of other substances on its own body, each monad constitutes a unique perspective on the world.[18] Through the organs of its own body, every monadic soul perceives the entire world from its own 'point of view', reflecting like a mirror the harmoniously ordered and varied whole: 'souls in general are living mirrors or images of the universe of created things' (*Monadology* §83: GP VI 621; M 31). Every monad is thus a unique representation of the infinitely varied multiplicity of monads with bodies (i.e. living creatures) that make up this world.

Crucially, this means that every monad is itself a thing of great beauty. Every monad is a 'unity-in-variety' and each of its passing perceptions is also a 'unity-in-variety', being a unified representation – from its own unique point of view – of the current state of the whole of this beautifully varied and perfectly ordered harmonious universe. As Leibniz conceives the matter, when we perceive perfections, the images of those perfections in our minds are repetitions of the perfections present in what is perceived and thereby serve to perfect the mind itself. In a discussion of pleasure in *On Wisdom*, he explains:

> *Pleasure* is the feeling of a perfection or an excellence, whether in ourselves or in something else. For the perfection of other beings is also agreeable, such as understanding, courage, and especially beauty in another human being, or in an animal or even in a lifeless creation, a painting or a work of

---

[18]Each monadic soul or entelechy perceives the effects of the external world on each of its body's parts, holding the entire causal sequence leading up to present states of each part in a single indivisible perception and making possible the corporeal substance's role as itself a cause of future effects on others (Phemister, 'Seeds and Souls', 137–40).

craftsmanship, as well. For the image of such perfection in others, impressed upon us, causes some of this perfection to be implanted and aroused within ourselves.

(*On Wisdom*: GP VII 86; L 425)

Leibniz's claim here that only 'some of the perfection' perceived is implanted in the perceiver should not be read as endorsing the notion that some monads do not perceive the whole universe. Leibniz is clear that monads always perceive the perfect whole.[19] Rather, the degree of perfection in the perceiver is proportional to the degree of distinctness of the monad's perceptions. Given that a monad's internal qualities are its perceptions and appetitions (the forces that move the monad from one perception onto the next in the sequence), and given that the perceptions can only be described in terms of their representational content together with the degree of confusion or distinctness of the representation, within the context of Leibniz's philosophical system, it makes sense to regard representations of external perfections as internal perfections of the mind, with the amount of the perfection 'implanted and aroused' in this way dependent upon how distinctly the external perfection is perceived. Consequently, although all monads are beautiful insofar as their perceptions echo internally the full range of variety and order of the external world, they are not all beautiful to the same degree, for the more distinctly the external perfection is perceived, the more perfect and beautiful is the perceiver.

Such qualitative differences in the degrees of perceptual distinctness give each monad its unique perspective on the world. This in turn allows each to make its own distinctive contribution to the harmonious variety and order of the world as a whole. The magnificent beauty of the world itself is manifested in the infinity of distinct and unique individuals as well as the order and harmony that comes from the fact that their perceptions all represent the same world, a world in which each finite individual mirrors not only the world, but also its Divine Creator:[20]

> every substance is like a complete world and like a mirror of God or of the whole universe, which each one expresses in its own way, somewhat as the same city is variously represented depending upon the different positions

---

[19] The general order of the universe demands that this be so (*Principles of Nature and Grace*, §§ 12–13: GP VI 603–4; M 275).

[20] The Leibnizian universe actually contains multiple harmonic orders: of bodies, of perceiving monads, and of souls and bodies united. Described more abstractly, these harmonies are systems of efficient causes and of final causes operating in parallel, each so finely tuned that, whatever is happening in the one corresponds exactly to what is happening in the other, in the manner of a mathematical bijective function. A further harmonic ordering, which perhaps need not concern us here, is the harmony between the kingdoms of nature and grace, where the kingdom of grace is the moral kingdom of God comprising all rational minds and spirits. For further details, see Phemister, 'Exploring Leibniz's Kingdoms' and Strickland, 'Leibniz's Harmony Between the Kingdoms of Nature and Grace'.

from which it is viewed. Thus the universe is in some way multiplied as many times as there are substances, and the glory of God is likewise multiplied by as many entirely different representations of his work.
(*Discourse on Metaphysics*, §9: GP IV 434; AG 42)

Variations in the degrees of perfection are essential if monads are to be distinguished both from each other and from God. On the other hand, of course, this means that all finite created beings contain some imperfections. Whether these imperfections are mere absences of perfection or actual imperfections, it is clear that in mirroring each other, each monad must represent not only the others' perfections, but also their imperfections. Just as the images of the perfections of others are 'implanted and aroused' in the perceivers, so too must be the images of others' imperfections and the question must be asked: Do these imperfections and their images pose a threat to the idea that Leibniz advances a positive aesthetics in respect of the natural world?

Superficially, this might appear to be the case, but Leibniz himself argues that any imperfections, taken in the context of the whole, benefit rather than detract from the beauty of the world. Obviously some imperfections are required in order to increase nature's variety and, in the case of less than fully distinct perceptions, these can be accommodated without any diminution of the order among things: order is retained because all monads perceive the same world in its entirety; maximum variety is introduced because the differing degrees of distinctness of monads' perceptions ensure that the world is perceived from all possible perspectives. In this way, therefore, what appears in a limited context an imperfection is, when considered in its wider context, a valued part of the whole. By way of illustration, Leibniz draws upon the visual and aural arts:

> Look at a very beautiful picture, and cover up except for some small part. What will it look like but some confused combination of colors, without delight, without art; indeed the more we examine it the more it will look that way. But as soon as the covering is removed, and you see the whole surface from an appropriate place, you will understand that what looked like accidental splotches on the canvas were made with consummate skill by the creator of the work. What the eyes discover in the painting, the ears discover in music. Indeed, the most distinguished masters of composition quite often mix dissonances with consonances in order to arouse the listener, and pierce him, as it were, so that, anxious about what is to happen, the listener might feel all the more pleasure when order is soon restored.[21]
> (*On the Ultimate Origination of Things*: GP VII 306; AG 153)

All created monads, insofar as they possess varying degrees of power, of knowledge or perception, and appetites or drive towards the good, are created in the image of a God conceived as omnipotent, omniscient and

---

[21] See also Leibniz to Sophie Charlotte, 9/19 May 1697: GP VII 545; LTS 160.

universally benevolent (Phemister, *Leibniz and the Environment*, Chapter 6). Leibniz's God is a fully rational God and in this respect, rational minds and spirits bear the closest resemblance to their Creator. The principles of reason are the same in God as they are in finite minds. The intelligible order and harmony that pleases God is the same order and harmony by which the world becomes intelligible to finite minds and that is pleasing to them when they contemplate it. In short, the principles of reason that govern human thought are believed by the Leibnizian theist to be exactly the same rational principles that characterize God's omniscience.

It is thus through theological belief in the existence of a rational and good God that we can be convinced that the order and harmony the natural sciences find in our multi-faceted world are not illusory, but that the world itself in all its parts is in principle intelligible. In the absence of a theological belief in divine rationality, the sciences cannot guarantee that the world itself conforms throughout to our intelligible reconstruction of it. Nor can they guarantee that our rational understanding of the world is an accurate representation of the world as it really is in itself. The natural sciences on their own cannot provide a sure guarantee that the principles of harmony and order that signal intelligibility to the human mind are applicable throughout all nature. The natural sciences have not, and perhaps never will, achieve complete understanding of the natural world. The major part remains hidden from view. Without the theological grounding jettisoned by Carlson, it remains a real possibility that natural scientific explanations indicate more about the nature of the human mind and its need to find order and harmony in what is essentially a chaotic system than it does about the actual nature of the world itself. For Leibniz's rational theist, however, the intelligibility of scientific explanation is justified by the theological belief in the rationality and intelligibility of the creator. To the Leibnizian believer, the order and harmony already discovered through the sciences is representative of a beautiful order and harmony throughout a world freely chosen by a rational God who finds goodness, perfection, and intelligible beauty in the maximization of variety within unifying orderliness.

CONCLUSION

In this paper, we have argued for a reading of Leibniz's aesthetics as an early example of a positive aesthetics of nature. In satisfying this modest aim we have also tried to show that Leibniz's version of positive aesthetics is plausible, not just in itself but also vis-a-vis modern versions of the doctrine, in particular that developed and endorsed by Allen Carlson. We have noted already Malcolm Budd's concern that the doctrine of positive aesthetics is often vague or imprecise, inasmuch as formulations of the doctrine typically leave it unclear whether it applies to nature as whole, or the biosphere, ecosystems, species, individual (natural) things, or individual (natural) events

(Budd, *Aesthetic Appreciation of Nature*, 97). As we have already shown, while the charge of vagueness or imprecision does apply to Carlson's version of the doctrine, it does not to Leibniz's; for whereas Carlson is not clear as to what exactly he thinks the doctrine applies, Leibniz explicitly tells us that the whole is beautiful, as is each and every component of that whole, that is, each monad together with its organic body, as well as every aggregate body composed of these. In this way, Leibniz's positive aesthetics is all-encompassing, and this too makes it more attractive than Carlson's: for when faced with the objection that the natural world contains instances of ugliness, for example, in malformed living things, Carlson swiftly concedes the point, thus making his version of positive aesthetics limited in scope. Leibniz's version, meanwhile, recognizes no exceptions whatsoever; he tells us that if we think we have found an example of natural ugliness or disorder, it is because we are not considering it aright. Thus to two of the most common objections to positive aesthetics, Leibniz's version of the doctrine offers the more satisfying response. Leibniz's version of positive aesthetics has a further advantage over Carlson's on account of its justification. While Leibniz, like Carlson, offered a justification of the doctrine based on the natural sciences, Leibniz realized that this alone was not sufficient. He saw that it needs to be supplemented by a rational theology that validates scientific method and that also brings into focus the contributions of the monads to the overall beauty of the created world. To be fair, a theological justification alone is also not sufficient. Although in this paper we have highlighted the deficiencies of the sciences in this regard, it is equally the case that without the natural sciences, we would be ignorant of the 'wonders of nature' that, for Leibniz, not only evidence the beauty of nature, but point further to the beauty of their creator and that underpin the beauty of all who perceive them. Together, however, the natural sciences and theology offer the strongest justification of a positive aesthetics of nature.

BIBLIOGRAPHY

A = Leibniz, Gottfried Wilhelm. *Sämtliche Schriften und Briefe*, edited by Berlin-Brandenburgische Akademie der Wissenschaften. Berlin: Akademie. Multiple volumes in 8 series, cited by series (reihe) and volume (band), 1923–.

AG = Leibniz, Gottfried Wilhelm. *G. W. Leibniz: Philosophical Essays*. Edited and translated by Roger Ariew and Daniel Garber. Indianapolis, IN: Hackett, 1989.

Barnes, Jonathan, ed. and trans. *The Complete Works of Aristotle*, 2 vols. Princeton, NJ: Princeton University Press, 1984.

Beiser, Frederick C. *Diotima's Children: German Aesthetic Rationalism from Leibniz to Lessing*. Oxford: Oxford University Press, 2009.

Berkeley, George. *Philosophical Writings*, edited by Desmond M. Clarke. Cambridge: Cambridge University Press, 2008.

Borruso, Silvano, trans. *St. Augustine: On Order (De Ordine)*. South Bend, IN: St. Augustine's Press, 2007.

Brady, Emily. 'Ugliness and Nature'. *Enrahonar* 45 (2010): 27–40.

Budd, Malcolm. *The Aesthetic Appreciation of Nature*. Oxford: Clarendon Press, 2002.

Burnet, Thomas. *The Sacred Theory of the Earth*. London: R. Norton, 1684.

Carlson, Allen. 'Nature and Positive Aesthetics'. *Environmental Ethics* 6, no. 1 (1984): 5–34.

Carlson, Allen. *Aesthetics and the Environment: The Appreciation of Nature, Art and Architecture*. London: Routledge, 2000.

Carlson, Allen. 'Hargrove, Positive Aesthetics, and Indifferent Creativity'. *Philosophy and Geography* 5, no. 2 (2002): 224–34.

Cassirer, Ernst. *The Philosophy of the Enlightenment*. Translated by Fritz C. A. Koelln and James P. Pettegrove. Princeton, NJ: Princeton University Press, 2009 [1951].

Fisher, John Andrew. 'Aesthetics'. In *A Companion to Environmental Philosophy*, edited by Dale Jamieson, 264–76. Oxford: Blackwell, 2001.

GP = Leibniz, Gottfried Wilhelm. (1875–90). *Die Philosophischen Schriften von Gottfried Wilhelm Leibniz*, edited by C. I. Gerhardt, 7 vols. Berlin: Weidman. Reprint, Hildesheim: Olms, 1965.

Gr = Leibniz, Gottfried Wilhelm. *Textes inédits*, edited by Gaston Grua, 2 vols. Paris: Presses Universitaires de France, 1948.

Guyer, Paul. *A History of Modern Aesthetics*, 3 vols. Cambridge: Cambridge University Press, 2014.

L = Leibniz, Gottfried Wilhelm. *Philosophical Papers and Letters*. Edited and translated by Leroy E. Loemker. 2nd ed. Dordrecht, Holland: D. Reidel, 1969.

Leibniz, Gottfried Wilhelm. *Theodicy: Essays on the Goodness of God, the Freedom of Man, and the Origin of Evil*. Translated by E. M. Huggard. La Salle, IL: Open Court, 1985.

LGR = Leibniz, Gottfried Wilhelm. *Leibniz on God and Religion*. Edited and translated by Lloyd Strickland. London: Bloomsbury, 2016.

LTS = Leibniz, Gottfried Wilhelm. *Leibniz and the Two Sophies: The Philosophical Correspondence*. Edited and translated by Lloyd Strickland. Toronto: Centre for Reformation and Renaissance Studies, 2011.

M = Leibniz, Gottfried Wilhelm. *Leibniz's 'Monadology': A New Translation and Guide*. Edited and translated by Lloyd Strickland. Edinburgh: Edinburgh University Press, 2014.

Moltmann, Jürgen. *God in Creation: An Ecological Doctrine of Creation. The Gifford Lectures, 1984-1985*. Translated by Margaret Kohl. London: SCM Press, 1985.

NE = Leibniz, Gottfried Wilhelm. (1982, repr. 1985). *New Essays on Human Understanding*. Edited and translated by Peter Remnant and Jonathan Bennett. Cambridge: Cambridge University Press.

Parsons, Glenn, and Allen Carlson. *Functional Beauty*. Oxford: Oxford University Press, 2008.

Phemister, Pauline. 'Exploring Leibniz's Kingdoms: A Philosophical Analysis of Nature and Grace'. *Ecotheology* 7, no. 2 (2003): 126–45.

Phemister, Pauline. *Leibniz and the Natural World: Activity, Passivity and Corporeal Substances in Leibniz's Philosophy*. Dordrecht: Springer, 2005.

Phemister, Pauline. 'Seeds and Souls'. In *Leibniz's Metaphysics and Adoption of Substantial Forms*, edited by Adrian Nita, 125–41. Dordrecht: Springer, 2015.

Phemister, Pauline. *Leibniz and the Environment*. London: Routledge, 2016.

Ray, John. *The Wisdom of God Manifested in the Works of Creation*. London, 1691.

SLT = Leibniz, Gottfried Wilhelm. *The Shorter Leibniz Texts: A Collection of New Translations*. Edited and translated by Lloyd Strickland. London: Continuum, 2006.

Strickland, Lloyd. 'Leibniz's Harmony between the Kingdoms of Nature and Grace'. *Archiv fur Geschichte der Philosophie* 2016.

Index

Absolute 5, 66; Leibniz's theory of God and 68; Plato on 68; Schellingian 77; selves as an 'organ' of 131
Absolute consciousness 129
'Absolute idealism' 129, 138n21; *see also* British idealism
Academie des sciences morales et politiques 85–7, 86n3, 88n7, 95, 97n15, 98n16
Académie impériale de médecine 101
Academy of Clermont-Ferrand 87
Academy of Lyon 93
'accidental views' *(zufällige Ansichten)* 45
'actual occasions' 120
Adams, Robert 161
Adventures of Ideas (Whitehead) 106, 119
Aesthetics and the Environment (Carlson) 193–4
aether 58–60; Bolzano's monadology 58; density of 58
agrégation 64–5, 87
Allgemeine Metaphysik (Herbart) 34–5, 37, 37n5, 40–2, 44–50
'allusion' 107n6
'alogical incompatibility' 173
Ampère, André-Marie 52
animism 85–6; Cousin on 94, 97; Lemoine on 99, 101; medical 93–5; philosophical 93–5; Scholastic 6
Animisme et vitalisme (Laprade) 93
anti-materialist reasoning 108–9
'Aphorismen zur Physik' ('Aphorisms on Physics'), (Bolzano) 53
appearance *vs.* reality 136–8
appetition 157, 184; described 111; as founded in drive 160–1; monads and 144, 207
Aristotelian Society 132
Aristotle 54, 75
Arnauld, Antoine 1

Arthur, Richard T. W. 143–4, 171
Athanasia, or Reasons for the Immortality of the Soul. A Book for Every Educated Person Who Wishes to be Reassured about This (Bolzano) 53
'Atomenlehre des sel. Bolzano' ('Atom Theory of the late Bolzano') (Příhonský) 53
atoms: Bolzano's monadology 55–7; indestructible 56; interaction of 56–7

Bacon, Francis 172
Baumgarten, Alexander Gottlieb 3, 9, 198–9
Beiser, Frederick 198
Benjamin, Walter 172
Berg, Jan 53, 60
Bergson, Henri 65, 172
Berkeley, George 197
best possible world, Leibniz vision of 205–9
Bilfinger, Georg Bernhard 3
Biran, Maine de 63, 65–6, 68–73; animism and 94; death of 88n6; duo-dynamism of 88n6; on French universities 88n6; introspection and substances 76–7
bodies: Bolzano's monadology 58–9; composite 1; Leibnizian 142; organic 2
Boirac, Émile 71
Bolzano, Bernard 5–6, 62; atoms and 55–7; on dominant atoms 57–8; existence of substances 54–5; metaphysics of 52–3, 60; polymathic talents 52
Bolzano's monadology 52–60; adherence and attribute 53–4; aether 58; atoms 55–6; bodies 58–9; criticism 59–60; dominance 57–8; interaction of atoms 56–7; motion and other change 58; overview 52–3; sources 53; substances 54–5
Bosanquet, Bernard 131

INDEX

Boscovich, Roger Joseph 56–7
Bouillier, Francisque 6, 85; Cousinian constraints on *Du principe vital et de l'âme pensante* 95–103; Cousinian spiritualism and 84–104; 1839 competition on history of cartesianism 85–8; *Histoire de la philosophie cartésienne* 88–93; return of animism on philosophical and medical scene 93–5
Boutroux, Émile 65n4, 71
Bradley, F. H. 129, 131
Brief Demonstration 151
'British Hegelianism' *see* British idealism
British idealism 129–32; *see also* 'Absolute idealism'
British idealist monadologies: Oakeley and 129–32; similarities with Leibniz's *Monadology* 129–30
Broad, C. D. 139
Bruno, Giordano 108
Burnet, Thomas 197

Carlson, Allen 9, 192–3; 'natural environment model' 194; view of positive aesthetics 193–8
Carr, H. Wildon 7, 130; Oakeley and 132; philosophy on monad 130
Cartesianism 69; abusive organicism and 96; Bouillier on 87; 1839 competition on history of 85–8
Cartesian mechanism 6, 96, 101
Cartesian occasionalism 93
Cartesian philosophy 86–7, 95
Cassirer, Ernst 198
causation: 'immanent' 121; 'transeunt' 121
Cavalieri, Bonaventura 13
Charlotte, Sophie 154, 203
Christianity 196–7
classical metaphysics 42
cognitive dualism 11, 13, 20, 29
Como Somos per Consello 78, 80
'compossibility' 173
conception of substance 147–53
concept of becoming 112
concept of change 112
The Concept of Nature (Whitehead) 114
concept of perception *vs.* prehension concept 121
'Concerning the Progress of Philosophy' (Maimon) 13, 20n6
Confessio philosophi (Leibniz) 174
Conspectus libelli elementorum physicae (Leibniz) 176
'continental philosophy' 149

Copernicus, Nicolas 202
Copleston, Frederick 36
corporeal substances 2, 111n8
Council for Public Instruction 85
Cousin, Victor 63, 66–73, 88
Cousinian eclecticism 85
Cousinian spiritualism 87; Francisque Bouillier and 84–104; institutionalized 87
Couturat, Louis 107
Cover, J. A. 142
Crane, Tim 131
'creative history of philosophy' 150, 165–6
A Critical Exposition of the Philosophy of Leibniz (Russell) 107, 114
critical monadology 5, 35
Critique of Pure Reason (Kant) 4, 12–13, 64
curve(s): folding paradigm and 171; function and 176; infinitely folded 179

Damiron, Jean-Philibert 86, 88, 97
Definitiones cogitationesque metaphysicae (Leibniz) 176
Degérando, Joseph Marie 67n6, 86
Deleuze, Gilles 8, 165, 170; comparative analyses of Spinoza and Leibniz 172; comparing world to infinitely folded curve for Leibniz 176–8; criticism of Leibniz 172–3; 'monadological metaphysics' 172–4; perspectivism on 186; physical world and 182; on 'regimes of representation' 172–3
De l'habitude (Ravaisson) 65, 71; analysis of habit 77–81; Biranian introspection and substances 76–7; metaphysics of monads 77–81; monadism in 71–81; objective and subjective analyses 72–6
de Risi, Vincenzo 64
derivative forces 77, 180
Descartes 7, 36, 39, 64, 66, 69–71, 84–103
De Stahl et de sa doctrine médicale (Lasègue) 99
'dialogical history' 8, 149–50, 164–6
Différence et répétition (Deleuze) 172–3
differential calculus 180–2
'differential quotient' 180
'differential unconscious' 188
Discourse on Metaphysics (Leibniz) 117, 151, 180, 184
dominance: atoms and 57–8; Bolzano's monadology 57–8
Doppler, Christian 58, 59
Drang see drive
drive: character of 156; force as 160; Heidegger account of 155–8; and points

214

INDEX

of view 158; self-surpassing 156; unifies a manifold 156; unifying function of 162–3
Duffy, Simon 171
Du principe vital et de l'âme pensante, ou Examen des diverses doctrines médicales et psychologiques sur les rapports de l'âme et de la vie (Bouillier) 85, 87, 95–103, 100
Du principe vital ou de l'âme pensante (Bouillier) 85
Dutens, Louis 3
'dynamism' 100, 160

eclipse of speculative reason 107–10
École Normale Supérieure 85–7
ego 38, 48; absolute 48; Fichtean 37; non-ego and 48; self-positing 48
empirical intuitions 14, 18, 23–4
empirical scepticism 18
Épicure opposé à Descartes (Rochoux) 86
Erdmann, Johann 36
Essais de théodicée (Leibniz) 174
Essai sur la philosophie en France au XVIIe siècle (Damiron) 97
Essay on Transcendental Philosophy (Maimon) 11, 13–27
essentialism 187
events 179–82; experiential 120; 'ideal' 182; mental 109; 'ordinary' 177; 'remarkable' 177
Examination of McTaggart's Philosophy II (Broad) 143
'exegetical history' 148; critique by Heidegger 149; *vs.* philosophical history 163
experience 116
experiential events 120
Exposition de la doctrine philosophique de Leibniz (Cousin) 68–9, 88n6, 98n17
expression: clear and obscure zones of 187–8; monad and 186–9

Faculty of Letters in Lyon 95
Faraday, Michael 52
Fenves, Peter 172
Fichant, Michel 173
Fichte, Johann Gottlieb 5, 21, 21n7, 34, 37–8, 48, 50, 62
Fichtean ego 37
Fischer, Kuno 36
Flügel, Otto 46
Fodor, Jerry 165
fold: defined 180; in Leibniz's texts 174–6; of matter 178–9; of the world 176–8

The Fold (Deleuze) 165
'fold going to infinity' 179
Fragments de philosophie cartésienne (Cousin) 92
French Academy 85
French philosophy 63, 66, 70n8, 98
French 'spiritualism' 85
'fulgurations': defined 133n12; monads, of God 133–4
function: curve(s) and 176; unifying, of drive 157, 162–3

Garber, Daniel 159, 161, 173
'General Overview of the Most Recent Philosophical Literature' (Schelling) 12–13, 28
Geschichte der Philosophie (Hegel) 36, 41n9
Gilbert, William 202
God: Leibniz and 113, 199–202; monads and 119n18, 208–9; natural sciences and 204; pre-established world and 182–3; theological belief and 209
God's mind 182–3
Green, T. H. 131
Grundsatzphilosophie (Reinhold) 28
Guyer, Paul 198

habit 63–81; analysis of 77–81; bad 79; in vegetal life 73
Hamilton, William 72
'hard' problem of consciousness 109
Harman, Gilbert 165–6
Hartz, Glenn 142
Harvey, William 202
Hauptpuncte der Metaphysik (Herbart) 37, 49
Hegel, G. W. F. 5, 34–7, 62, 66, 68
Heidegger, Martin: account of drive 155–8; being of other beings concept and 154; on Cartesian conception of body 151; 'creative history of philosophy' 150; critique of exegetical history 149; 'dialogical history' 149–50; on Leibniz's conception of substance 147–53; on Scholastic power 153; understanding of 'monad' 151, 152
Herbart, Johann Friedrich 5, 34; as direct realist 48–9; as indirect realist 49; as scientific realist 49
Herbart's monadology 34–50; failures of past metaphysics 36–41; historical significance 34–6; outlines of 42–6; realism 46–50

INDEX

'Herbart's Monadology' (Beiser) 5
Histoire de la philosophie cartésienne (Bouillier) 85, 87, 88–93
Histoire et critique de la revolution cartésienne (Bouillier) 88
Hoffmann, Anna 56
Hölderlin, Friedrich 35, 37, 37n7
Hume, David 13–15
Hume Variations (Fodor) 165
hylomorphism 91

'ideal events' 182
idealism 5, 35, 47; absolute 5; British 129–32; Fichte on 37; Herbart on 47–9; Leibnizian 131; McTaggart's views on 130; panpsychistic 111; personal 129; transcendental 5, 26, 49
ideal space 46
Ideas for a Philosophy of Nature (Schelling) 28
'immanent' causation 121
immanentist pantheism 80
inclusion: Leibnizian notion of 184; in the monads 182–5
Individuals (Strawson) 150
Influence de l'habitude sur la faculté de penser 68
'influx' 121
institutionalized spiritualism 99
intellectual intuition 20–1, 26, 38
Introduction à la philosophie médicale (Laprade) 93
intuition: empirical 14, 18, 23–4; intellectual 20–1, 26, 38

Jouffroy, M. 86, 88n6, 95
'judgement of experience' 15

Kant, Immanuel 4–6, 12–28, 32, 35–40, 47, 49, 52, 56
Kantian cognitive dualism 11–32
Kantianism 12
Kantian problems 13–18
Kantian transcendental philosophy 18
Kepler, Johannes 202
Knecht, Herbert 172
Kroner, Richard 36

La Logique de Leibniz (Couturat) 107
La Philosophie en France au XIXe siècle (Ravaisson) 64
Laprade, Richard de 93–4
Lærke, Mogens 8
Lasègue, Ernest-Charles 99

Lehrbuch zur Einleitung in die Philosophie (Herbart) 38n8, 41n10, 43, 47
Leibniz (Arthur) 171
Leibniz, Gottfried Wilhelm: on Cartesian physics 179–80; conception of substance 147–53; critique of materialism 109; death of 3; differential calculus *see* differential calculus; ego as guiding clue for 153–5; mannerism 187; Mill-Argument 108–9; *Monadology* 1, 3, 11, 74, 80, 87, 91, 107–8, 129, 133, 172, 187–8; on nature of causation 117–19; positive aesthetics 193, 198–201; pre-established harmony 113–14; Principle of Sufficient Reason 108, 109–11, 204; rationalist method 4; sciences and 201–5; vision of best possible world 205–9
Leibniz: Body, Substance, Monad (Garber) 159, 161
Leibnizianism 11, 64
Leibnizian monadology 11–32
Leibnizian monads 141–3; *see also* monad(s)
Leibnizian notion of inclusion 184; *see also* inclusion
Leibnizian spiritualisms 66–71; Maine de Biran and 66–71; Victor Cousin and 66–71
Leibniz's Metaphysics (Mercer) 171
Lemoine, Albert 98, 99
Le Pli. Leibniz et le Baroque (Deleuze) 8, 170–1, 172
Le Système de Leibniz et ses modèles mathématiques (Serres) 174
Le Vitalisme et l'animisme de Stahl (Lemoine) 99
L'Image-movement (Deleuze) 172
L'Image-temps (Deleuze) 172
Locke, John 102, 109
Lodge, Paul 8
Logique du sens (Deleuze) 173, 181–2

Madinier, Gabriel 72
Maimon, Solomon 4–5, 13, 19n4; 'intellectual intuition' and 21; Leibnizian solutions of 18–27; 'objective reality' and 14; on *'Quid juris?* 17
Malebranche, Nicolas 71, 84, 93, 98
manifold, of monad: as character of drive 156; drives as unification force for 156; nature of 157–8
mannerism 187
material bodies 58–9; *see also* bodies

INDEX

materialism 90, 93–4, 99, 101, 109; Leibniz–Whitehead critique of 109; Principle of Sufficient Reason and 111
materialist monism 90
matter: fold of 178–9; *Monadology* and 178–9
McDonough, Jeffrey K. 142
McTaggart, J. M. E. 128–31; Oakeley critique of 136–41; on selves 138; unreality of time 128, 136–7, 140
McTaggart, John McTaggart Ellis 7
mental events 109
mentalism 111
Mercer, Christia 171
Metaphysical Foundations of Logic (Lodge) 8
The Metaphysical Foundations of Logic (MFL) 147, 150–1
metaphysics: failures of past 36–41; monadological 8; of monads 77–81
Metaphysics (Baumgarten) 3
metaphysics of substance: Russell on 114–17; Whitehead on 114–17
Mill-Argument 108–9
Modes of Thought (Whitehead) 117
Moltmann, Jürgen 197
monad(s) 63–81; appetition and 144; Carr's view of 130; described 110; expression and 186–9; as 'fulgurations' of God 133–4; Heidegger's analysis of 152, 161; inclusion in 182–5; individuality of 187; Leibniz's description of 111–13; metaphysics of 77–81; nature of 160; plurality of 1; prehension and 119–23; as souls 111; spiritualism with 84–104; as subjects 184; substance as 150–3, 206; as substrata 114; unity of 152
monadism 115–16; in *De l'habitude* 71–81; Leibnizian 63–5, 115; Monism and 116
'monadological metaphysics' 8, 172–4
Monadology (Leibniz) 1, 3, 11, 74, 80, 87, 91, 107–8, 129, 133, 172, 174–5, 187–8; creation of monads 133; Deleuze's approach to 189; idealisms in 129–30; matter and 178; Serres on 174; soul in 187
Monism 115
motion: Bolzano's monadology 58; Herbart on 45–6, 49

Nagel, Thomas 124n24
Napoleon III 64
'natural environment model' 194
nature of causation: Leibniz on 117–19; Whitehead on 117–19

Nature of Existence (McTaggart) 130, 137
Negotium otiosum ('A Tiresome Quarrel') (Stahl) 98
New Essays (Leibniz) 3
'normal' lines 187
Nouveaux essais sur l'entendement humain 102
Nouvelle monadologie (Renouvier) 64
Nouvelles études sur le spiritualisme (Pidoux) 101

Oakeley, Hilda 7, 128–9; background 132; British idealist monadologies and 129–32; Carr and 132; and Carr's philosophy on monads 130; critique of McTaggart 136–41; *vs.* Leibniz views on reality of time 141–4; reality of time and 133–6; on selves 135
Obereit, Jacob Hermann 27
'objective appearance' 45, 50
occasionalism 91–3, 175
Oeuvres philosophiques latines & françoises de feu Mr. de Leibnitz (Raspe) 3
On Nature Itself (Leibniz) 70, 76
'On the Corrections of Metaphysics and the Concept of Substance' 70
On the Emendation of First Philosophy and the Notion of Substance (Leibniz) 152, 159
'On What is Independent of Sense and Matter' 154
On Wisdom (Leibniz) 206
Opera Omnia... nunc primum collecta, in classes distributa, praefationibus et indicibus exornata (Dutens) 3
'ordinary' events 177

Pacidius philalethi (Leibniz) 174, 175
'pan-metaphysics' 86
panpsychistic idealism 111
pantheism 84, 88–93, 91; immanentist 80; Spinozistic 6
Paradoxien des Unendlichen (Paradoxes of the Infinite) (Bolzano) 53, 60
past metaphysics, failures of 36–41
Pathmarks (Heidegger) 147–51, 153
Peirce, Charles Sanders 172
perception 157; as founded in drive 160–1
personal idealism 129
Personal Idealism (Sturt) 129
personality: metaphysics of 69; Oakeley's definition of 133
perspectivism 185–6

INDEX

Phemister, Pauline 9
'philosophical history' 148–9; vs. 'exegetical history' 163; legitimacy of 165–6
Philosophical Papers and Letters (Leibniz) 159
Philosophie contemporaine (Ravaisson) 65, 72–3, 72n11, 76, 78–80
Philosophische Parallelen 28
Philosophy of Personality (Oakeley) 132, 133–6
Physical Monadology (Kant) 4
Pidoux, Claude 101
Plato 35, 68
pluralism 44
positive aesthetics 192; Carlson's view of 193–8; Leibniz view of 193, 198–201; 'problem of evil' and 196; 'problem of ugliness' and 196; science and 195–6; theological justification of 196–7
potentia nuda 152
prehension concept 119–23; vs. concept of 'perception 121; described 121
Příhonský, Franz 53
primitive active force 77
primitive passive force 77
Principes de la nature et de la grâce (Leibniz) 174, 176, 187–8
Principia Mathematica (Whitehead and Russell) 115
Principle of Sufficient Reason 108, 109–10, 204; materialism and 111
Principles of Nature and Grace (Leibniz) 117
Principles of Psychology (James) 120
Pringle-Pattison, Andrew Seth 129, 131; on soul-substance doctrine 131
Process and Reality (Whitehead) 106, 110, 113, 115
Protogaea (Leibniz) 175
Prussian Royal Academy of Sciences 11

'quantity of motion' 151
Quinet, Edgar 66

Rapport sur le concours sur la philosophie de Leibniz (Damiron) 97
Raspe, Rudolf Erich 3
rational dogmatism 18
rational psychology 21
rational spiritualism 85
Ravaisson, Félix: on metaphysics 66; theory of substance 63–81
Ray, John 197

realism 47; Herbart's 46–50
reality 43; vs. appearance 136–8; basic principles of 110–11; Leibnizian theory of 124
reality of time: Leibniz and 141–4; Oakeley and 133–6
The Realm of Ends or Pluralism and Theism (Ward) 130
'real phenomena' 182
real space 46
reciprocal determination 24
'Reconciling Leibnizian Monadology and Kantian Criticism' (Fincham) 4
'regimes of representation' 172–3
Reid, Thomas 72
Reinhold, Karl Leonhard 28, 37–8
'remarkable' events 177
Renouvier, Charles 64, 71
Rescher, Nicholas 142
'retention' concept 120–1; metaphysical interpretation of 123
Rochoux, Jean-André 86
romanticism 35
Rosenkranz, Karl 36
Royal Bohemian Society of Sciences 53
Russell, Bertrand 7, 55, 60, 107, 131; on Leibniz's views on time 143; on metaphysics of substance 114–17

Saint Augustin 176
Schelling, Friedrich Wilhelm Joseph 5, 12, 27–32
Schlegel, Friedrich 35
Scholastic animism 6
Scholastic power 152–3
Schwab, Johann Christoph 12
Science and the Modern World (Whitehead) 106, 107, 108
sciences: Leibniz and 201–5; positive aesthetics and 195–6
Scottish common sense school 66
'Second Antinomy' 11
'self-preservation' *(Selbsterhaltung)* 40
selves: McTaggart on 138; Oakeley on 135
Serres, Michel 174
Simon, Jules 66
Simons, Peter 6
Sleigh, Robert, Jr. 148; 'philosophical history' 148–9
Smith, Daniel W. 171
Socrates 54
Spinoza, B.: Monism and 115; single encompassing reality theory 114

INDEX

Spinoza et le problème de l'expression (Deleuze) 172
Spinozistic pantheism 6
spiritualism 68; Cousinian 84–104; French 85; institutionalized 99; Leibnizian 66–71; with monads 84–104; rational 85
'spiritualist positivism' 64
Stahl, Georg Ernst 7, 98
Stahl et l'animisme (Lemoine) 99
'static ontological genesis' 173
Stewart, Dugald 72
St. Hilaire 86
Strawson, P. F. 150
Strickland, Lloyd 9
Sturt, H. G. 129
substance: Bolzano's monadology 53–5; conception of 147–53; as monad 150–3, 206
substratum 120
Système nouveau (Leibniz) 185

teleology 38
'temporal passage' 135
temporal 'perceptions 137
theory of actual occasions 7
theory of substance 63–81
time: Leibniz description of 141–2; as 'phenomenon bene fundatum' 137; Rescher on 142
'Time and the Self in McTaggart's System' 138
Traité de l'Homme 96
transcendental idealism 5, 26, 49; *see also* idealism
transcendental realism 26
'transeunt' causation 121

'unity-in-variety' 198, 206
unity of monads: Heidegger on 151; primordial status 152; unifying function of drive and 162–3
'Universal Law of Being' 74
unreality of time 128, 136–7, 140

Van Helmont, Jan Baptist 97
van Leeuwenhoek, Antonie 202
Vermeren, P. 68
Versuch einer neuen Logik oder Theorie der Denkens (Maimon) 28
'vicediction' 173
vis 160
vis activa 152–3
vitalism 85
Vom Ich als Princip (Schelling) 28

Wallace, William 131
Ward, James 7, 130
Whitehead, Alfred North 7; anti-materialist reasoning 108–9; background 107; critique of materialism 109; on eclipse of speculative reason 107–10; on Leibniz's pre-established harmony 113–14; metaphysical works 106; on metaphysics of substance 114–17; on nature of causation 117–19; on prehension concept 119–23; on 'retention' concept 120–1; on scientific theories 111
Williams, D. C. 135
Wissenschaftslehre (Bolzano) 48, 52
Wolff, Christian 3, 198–9
world: Deleuze on physical 182; God's mind and 182–3; Leibniz and pre-established 182–3; Leibniz vision of best possible 205–9